Great Powers and Geopolitical Change

Great Powers and Geopolitical Change

JAKUB J. GRYGIEL

The Johns Hopkins University Press

Baltimore

© 2006 The Johns Hopkins University Press
All rights reserved. Published 2006
Printed in the United States of America on acid-free paper

Johns Hopkins Paperback edition, 2011
2 4 6 8 9 7 5 3 1

The Johns Hopkins University Press
2715 North Charles Street
Baltimore, Maryland 21218-4363
www.press.jhu.edu

The Library of Congress has catalogued the hardcover edition
of this book as follows:
Grygiel, Jakub J., 1972–
Great powers and geopolitical change / Jakub J. Grygiel.
p.   cm.
Includes bibliographical references and index.
ISBN 0-8018-8480-2 (hardcover : alk. paper)
1. Geopolitics—History. 2. Political geography—History. I. Title.
JC319.G78 2006
327.1′01—dc22
2006005256

A catalog record for this book is available from the British Library.

ISBN 13: 978-1-4214-0415-8
ISBN 10: 1-4214-0415-X

CONTENTS

Like most books, this one is the product of many minds that offered ideas but also encouragement, criticism, and stylistic advice. The list is long, and the help received from each individual unique; hence the following are not necessarily named in order of importance. I owe an enormous debt of gratitude to my adviser at Princeton, Aaron Friedberg, who was a constant source of advice and support. The members of my dissertation committee, Michael Doyle, Wolfgang Danspeckgruber, and Kathleen McNamara, made valuable comments on papers on similar topics, on individual chapters, and on the entire manuscript. Colleagues and friends at Princeton, William Block, Colin Dueck, Lyle Goldstein, and Daniel Markey, followed the development of this project from its early stages and commented on parts of the manuscript. William Block had the patience to read the entire manuscript. Dianne Dobbeck gave me detailed comments on chapter 4 (Venice), paying particular attention to style. Jim and Masako Shinn supplied me with a perfect environment, their house in Princeton, in which to work on a large part of the manuscript. Andrew Rhodes, Francisco de Santibanes, and Erin Schenck provided indispensable research assistance toward the end of my writing.

At the Johns Hopkins University Press, Henry Tom supported this book and guided it through the entire publication process. Claire McCabe and Juliana McCarthy were also helpful in bringing the manuscript to its book form. Joanne Allen painstakingly copyedited the manuscript, clarifying the argument and tightening the style.

Most of this book was written while I was at Princeton University, and my research was made possible by several grants. My PhD studies were supported by a grant from the graduate school, which were followed by a fellowship from the Woodrow Wilson Society of Fellows. The Smith Richardson Foundation, through the Research Program in International Security, directed by Aaron Friedberg, generously funded parts of my research. I also benefited from a Bradley Foundation postdoctoral fellowship at Princeton.

A leave of absence funded by a junior-faculty research grant from the Smith Richardson Foundation allowed me to finish the manuscript. Finally, some of the final revisions, including the last chapter, were made possible by a summer research grant from the Johns Hopkins University School of Advanced International Studies.

My parents, to whom this book is dedicated, supported my intellectual journey in many ways, and this book is only one, and perhaps the least important, fruit of their love and education.

Finally, my wife, Priya, followed this project from its outset and spurred me to write better, avoiding complicated sentences that were dangerously close to Latin grammar, and was the best cheerleader. Our son, Tobias, who arrived at the end, added a good dose of fun and, above all, put everything in perspective.

The subject of this book is geopolitics and geostrategy. It is structured around historical case studies of Venice, the Ottoman Empire, and Ming China from roughly the eleventh to the sixteenth century. Because both the subject and the choice of case studies are uncommon in most current international relations theory, it seems appropriate to begin by justifying them and clarifying some terms. The purpose of this book is threefold: to bring geography back to the study of international relations, to examine the importance of a foreign policy that reflects the underlying geopolitics, and finally to draw from history lessons for U.S. foreign policy.

For a variety of reasons that have little to do with reality and reflect mostly the mood swings of academic fashion, in the last four decades geography has played an increasingly smaller role in international relations theory, resulting in a growing gap between theory and practice. The objective of theory is the simplification of reality in order to increase the explanatory scope of the hypotheses. The resulting abstractions are of little value to those interested in acquiring, managing, and using political power in the international realm. I argue, however, that geography remains important even in this age of instant communications and of the apparently unlimited reach of American power. States wield their power not on a billiard table but on seas, through mountains, and across plains, and their success depends above all on their ability to match their foreign policy to the underlying geopolitical situation.

To clarify my argument, in chapter 2 I explain three variables: geography, geopolitics, and geostrategy. *Geography* is the geological reality of the earth, composed of mountains, rivers, seas, climate, and so on. *Geopolitics* is a combination of geological features (e.g., natural resources) with human activity (e.g., production and communication technology) that alters the value of places. While geography does not change (e.g., the location of oil fields is the same now as it was two hundred years ago), the value given to it can fluctuate dramatically with the development of new technologies (e.g., over the course of several decades the diesel engine increased the importance of oil-rich regions from the Caspian Sea to the Middle East). In short, geopolitics is defined by the location of natural and economic resources and the lines of communication linking them. It is a map of sorts, assigning strategic value to places. Finally, *geostrategy* describes where a state directs its military and diplomatic efforts. Because of limited resources, states, even the most powerful ones, must choose where to project their power and influence, imparting a clear geographic direction to their foreign policy.

My argument rests on the interaction between geopolitics and geostrategy. When states take into account the geopolitical situation and pursue a geostrategy that reflects it—that is, when they control centers of resources and lines of communications—they increase and maintain their position of power. When a state fails to, or chooses not to, extend its control over resources and routes, other states are likely to fill the vacuum. A state that commands resources and routes also has the ability to accumulate wealth and exercise leverage over others. It guarantees itself access to key resources, while at the same time it can deny, or even threaten to deny, similar access to other states.

Let me pause here for a moment and address at least one possible criticism of my argument. A connection between geopolitics, for instance, trade routes, and the political fate of a state seems self-evident, and consequently my argument might be considered as a truism. After all, states do not act in some abstract vacuum, but within the concrete geographic setting that is theirs thanks to geological chance and political events. This is a criticism that I find difficult to respond to simply because I agree with its premise, and my whole argument is founded on a similar understanding of international relations. It seems to me almost obvious that geography and geopolitics have a profound impact on the political fate of states, in particular states that do not recognize the need to control routes and resources lose power. What follows in these pages is simply an attempt to strengthen this argument that some may consider a truism.

These days the predominant view is that geography or geopolitics has little, if any, influence on the success or failure of states, especially in this modern, globalized age. This argument is based on the premise that control of certain geopolitical objectives has lost its significance in the age of globalization because states can prosper by tapping into the market for resources and access to routes, obviating the need to extend their command over them. The idea of *globalization,* broadly defined as the reality created by a growing volume and intensity of economic interactions among states, together with a parallel, gradual decline in the belief that states are constrained by geographic factors (the subject of chapter 1), underpins this argument. My goal in this book is to show that geography and geopolitics are important and that, as in the past, exclusive control over routes and resources is a source of power that cannot be replaced by a "market." In many ways, the purpose of the following pages is to restore the truism that geography is vital for the foreign policy of states. I do so by clarifying the relevant concepts—geography, geostrategy, geopolitics—and their relationship in chapter 2, and I illustrate them in the historical case studies.

Finally, I want to make clear at the outset that I do not propose a grand new theory of international relations or foreign policy centered on geographic variables. In fact, I am skeptical about the possibility of achieving such a "geopolitical theory," and I am aware that when such theories have been formulated, they have tended to fall into determinism or, worse, dangerous ideologies. The objectives of this book are more modest: I want to take the dust off some old-fashioned concepts and insights that have been abandoned by the majority of academic and policy writings. There is still use, especially in policymaking, as I show in the chapter 7, for an appreciation of geopolitics.

The second purpose of this book concerns policy. Specifically, I am interested in where power is applied most effectively or, alternatively, where states should apply power in order to increase or maintain their position. This question derives from two larger questions that characterize the art and study of politics, namely, how to obtain power and what to do with it. I do not propose to answer the questions how to generate power and wealth or why some states are more aggressive than others in gathering power. Rather, I am interested in examining the second question—what to do with power—and take the first one for granted. My question therefore can be phrased as follows: once a state has decided to increase its power and once it has power, what does it do with it?[1]

This is a vast question that encompasses questions on the "common good" or the very purpose of political life. Power can serve multiple interests, including its own. I limit myself to observing the multiplicity of reasons for which power is used, from religious motivations (early Ottomans) to commercial greed (Venice), without seeking to examine their origins in depth. I am interested in a very narrow part of the question concerning what to do with power, namely, the geostrategic use of power. Where does one use it? Why there? If one wants to succeed as a great power, for whatever reason, where should one direct one's attention and energies? I argue that there are some objective geographic goals, such as trade routes or resources (the geopolitics), that states must pursue in order to maintain or increase their relative strength. The study of such objectives, which are historically specific (i.e., the geostrategic objectives of fifteenth-century Venice are not the same as those of the United States in the twenty-first century), is the key to a foreign policy or geostrategy that brings strategic gains to a state. Conversely, the knowledge of these objectives offers some insights as to where a state is likely to expand. It will not enable one to predict whether a state will do so or not because, as mentioned above, a state might choose not to expand or might expand on the basis of ideological goals, leadership whims, or simply a mistake in reading the underlying geopolitics. That is why the knowledge of geopolitics indicates only the *likely* direction of a state's foreign policy.

Furthermore, because geopolitics does not determine the strategic direction of a state, my argument and the case studies I examine can be read as a story of missed opportunities. The fate of states, and especially of great powers and empires, rests in their own hands; they rise because of specific policies, and they decline because of particular decisions. Venice, the Ottoman Empire, and Ming China declined because they chose to pursue specific policies that in the end did not bring them strategic benefits and in fact put them at great disadvantage. This point was well put by Montesquieu, who wrote in *Considerations on the Causes of the Greatness of the Romans and Their Decline* that "it is not chance that rules the world. Ask the Romans, who had a continuous sequence of successes when they were guided by a certain plan, and an uninterrupted sequence of reverses when they followed another."[2] Neither chance nor geopolitics determines the course of history.

The third objective of this book is to derive from history policy lessons for the United States. More precisely, I am interested in how to maintain the American preponderance of power in the world. That is why the study of the historical cases is mercenary: I want to find policy wisdom that could be applied to current strategic challenges. Such interest has also driven the choice of the historical period and cases. The Venetians, Ottomans, and Chinese were all great powers that faced a dramatic change in geopolitics caused by the fifteenth-century discoveries of new routes and continents. The layout of routes and the location of resources changed, and the three powers had to adjust their geostrategies to the new geopolitical reality. Ultimately, their failure to maintain their position of strength was due to a combination of geopolitical change and mistaken geostrategies that focused on the strategically less vital regions (e.g., Venice expanded toward the Italian mainland rather than strengthening and expanding its maritime hegemony). The ensuing gap between the underlying geopolitics and the geostrategy doomed these powers.

This history has some parallels with the current situation faced by the United

States. Although it is still too early to tell, the United States might be confronting a geopolitical change on a scale similar to that faced by Venice, the Ottoman Empire, or Ming China. The growth of China in East Asia, combined with the collapse of the Soviet Union and the continued weakness of Russia in central Eurasia, is creating a geopolitical situation radically different from that of the previous several decades. Regions that until now were relegated to the periphery are becoming strategically important because of their growing economic wealth and potential military power. This change in the locale of power is also leading to a rearrangement of the most vital trade routes in the world: the sea lanes of East Asia might become the key lifelines in the world. They already supply Japan with 90 percent of its energy needs. Whoever controls these routes is likely to wield a power comparable to that of Venice in the fourteenth and fifteenth centuries.[3]

It is important to make two observations about the case studies and their relevance for the present. First, a strong argument can be made that the three great powers examined here lost their position of superiority because their internal systems failed rather than because they mismanaged their foreign policy. According to this argument, Venice, the Ottoman Empire, and Ming China were overcome by the more efficient nation-state model of Western (or Atlantic) Europe, which because of internal conditions (such as a Protestant ethic) or exogenous shocks (such as war or revolution) became the predominant form of state organization. Venice, for instance, simply could not compete with larger states such as France, which managed to harness national resources in a more effective way and to direct them to the outside to strengthen security and expand influence.[4] In brief, the three powers examined here became relics of a less efficient state model, and the mismatch between their geostrategy and geopolitics was not the decisive factor.

To a degree, this is a classic chicken-and-egg question. The decline of the three powers I present in the next chapters occurred while they were being challenged by new, energetic "nation-states," and it is plausible that they could not respond effectively to geopolitical changes because of an inefficient state model. However, I do not really engage in this argument for two reasons. First, the mismatch between their foreign policy and the underlying geopolitics, at least to the degree that it is possible to gauge it, predates the rise of the nation-states. For instance, Ming China withdrew its influence from the oceans before the arrival of the Portuguese, and a direct confrontation between western European powers and China never occurred, leaving us with one of the greatest "what if's" of history. Second, I do not address in great detail the internal causes of geostrategy. It may be the case that in some instances, perhaps in Venice and the Ottoman Empire, the inability to alter geostrategy (or the failure to implement fully a new geo-strategy more reflective of the changes in routes and centers of resources) was due to the weakness of the state model. My main purpose, however, is to point to the importance of this mismatch rather than to examine extensively its causes.

The possibility of multiple explanations for the events that characterize these case studies leads me to the second caveat concerning the relevance of history for current policy challenges. The three historical cases examined here should not be considered as unequivocal roadmaps for the present, in part because, as men-tioned above, some of the causes behind the foreign policies of these great powers may be different. It is also obvious that the technologies at their disposal were different from those the United States can harness today; the political rules gov-erning domestic as well as, to a degree, international relations were unique to that period; the configuration of geopolitics in the fourteenth to sixteenth centuries is not even slightly similar to that of the present. The list of differences is long, and the lessons we can learn from the past cannot be applied fully and uncritically to the present. Venice and Ming China were products of their time and space. However, history, it has been said, teaches by analogy, not by identity. These case studies illustrate certain challenges of matching geostrategy to the underlying geopolitics and, above all, the dangers stemming from a misalignment between the two. Therefore, the reader who seeks in the following pages a clear connection between the past great powers and the present-day United States, hoping to find the solution to the strategic challenges looming on the horizon, will be dis-appointed. The task of this book, and the lesson that it tries to convey, is more modest because ultimately it only suggests that now as in the past, great powers, irrespective of their material strength or domestic regime, must heed the geo-political reality in order to maintain their position.

The modesty of this claim, however, should not overshadow the importance of a geostrategy that reflects the underlying geopolitics, especially because in de-mocracies, and perhaps in the United States in particular, there is a certain reluc-tance to face the realities of power and its requirements. A growing faith in the power of globalization to overcome the limitations of geopolitics, and the need to control key locations and routes, exacerbates this tendency to discount the influ-ence of geopolitics. I am not the first to notice such a trend, and undoubtedly I am not the first to try to correct it.[5]

The book is organized as follows: In chapter 1 I trace the rise and decline of the study of geography in international relations and argue that the disappearance of geography and geopolitics has been induced more by academic trends than by a fundamental change in how strategic interactions among states occur. In chapter 2 I articulate my argument by defining three concepts—geography, geopolitics, and geostrategy—and how they relate to one another. In the succeeding chapters I illustrate my argument through three historical examples. In chapter 3 I describe the geopolitical situation in Eurasia in the sixteenth century. Then in chapters 4–6 I examine the geostrategies of three great powers—Venice, the Ottoman Empire, and Ming China—and the attempts to adapt them to changed geopolitical situations. Finally, in chapter 7 I draw from the case studies some lessons for the United States, focusing on the importance of adapting its geostrategy to a changing geopolitical environment characterized by a shift of power toward East Asia and the Pacific theater.

Great Powers and Geopolitical Change

CHAPTER ONE

The Premature Death of Geography

Is geography still important in the examination of foreign policy? While the prevailing literature of international relations theory discounts the role of geography, I argue that geography continues to be a key variable shaping the grand strategy of states.

The goal of this book is to bring geography back into the discussion of international relations.[1] I begin, in this chapter, by examining the rise and decline of the study of geography and politics, or geopolitics. This field began as a natural science and gradually moved into social science, resulting in a growing abstraction that privileged social concepts to the detriment of geography. The resulting theories, devoid of the complexity of geography, are elegant but detached from reality and incapable of offering tools for comprehending and formulating foreign policy. A revaluation of geography results perhaps in a more complex theory that, shedding some elegance and simplicity, can be more policy-relevant and more realistic.

Next I examine geography, the independent variable. I show that in order to avoid determinism, which characterizes so much of the geopolitical literature, we must understand geography as a combination of immutable geological facts (such as the patterns of lands, seas, rivers, mountain ranges, and climatic zones) and the human capacity to adapt to them through changes in production and communications technology. The result of this combination of geology and human activities is three variables: the layout of trade routes, the location of resources, and the nature of state borders. At the level of foreign policy, geography is a geopolitical reality to which states respond by formulating and pursuing a geostrategy.

Finally, I analyze the response of states to the geopolitical reality by examining three case studies: the Venetian and Ottoman empires and Ming China. I argue that the most successful states are those that match their geostrategy to the underlying geopolitical reality. States that protect their home territory (or, as a proxy, have stable borders), then pursue control of resources, and that secure routes connecting them with centers of resources, increase and maintain their relative power.

The Renaissance of Geography

Before examining the reasons for the current neglect of geography in international relations theory, it seems appropriate to place this project in the context of recent trends. In fact, there is a renaissance of geography. Specifically, there are two distinct trends. First, there is a renewed interest in geography in social sciences, in particular in international economics and political science. Second, although there has always been a strong interest in geography in analyses of foreign policy, such interest surges in moments of great geopolitical upheavals.

This project fits into a wider renaissance of geography. As I discuss later, geography has been reintroduced as an explanatory variable in economics and political science.[2] In part this is due to the desire to complicate theories of economic growth and state behavior, increasing their explanatory power. The drive to make theories more elegant and simple has led to the abandonment of complicated variables such as geography, which are not easy to categorize. Moreover, the misuse of geography by ideologically motivated authors, such as the Nazi theoretician Karl Haushofer, has given negative connotations to any analysis that tried to marry geography and politics.

However, the theories resulting from such a dislike of geography are too abstract to explain politics and offer useful policy advice. The acknowledgment of these limitations has led to the current renaissance of geography, which is still in its early stages and limited mostly to the policy world. For instance, there are no theories or systematic treatments of how geography influences the foreign policy of states. In fact, there is no agreement even on what *geography* means. The purpose of this book is to add to this literature by tracing the reasons for the rise and decline of geography in political science and by clarifying some of the most used concepts—geography, geopolitics, geostrategy—through a set of case studies.

The second context of this project is policy analysis. The study of geography and politics tends to become more important in times of great political changes. It is not by chance that Halford Mackinder's seminal book, *Democratic Ideals and Reality*, was first published in 1919, after World War I, and reprinted in 1942, during World War II. Dramatic changes in the geographic distribution of power require a geographic framework that offers tools to understand them. In order to redraw a map of power, one needs to know geography. But because such changes are rare, the appreciation of geography is cyclical and falls into academic oblivion during periods of international stability.

The geopolitical changes that followed the collapse of the Soviet empire spurred the current interest in geography. The "post–Cold War" multipolar international system has heightened the awareness of power differentials in the world. Power is concentrated in specific regions (e.g., North America and Western Europe), and how these regions relate to one another depends in large measure on where they are located.

Within such a policy context there are two analytically distinct questions. First, there is the geopolitical question how to map the new distribution of power. Which regions of the world are strategic? Where are the potential zones of conflict? The second question is more directly related to policy because it deals with the new geostrategic imperatives of states. Where should states concentrate their attention? Where is projection of power most effective and necessary?

Over the last decade the U.S. position of unrivaled power and almost unlimited reach led to a reevaluation of American geostrategic priorities. Instead of focusing on the European theater, the United States might have to retool its structure of military bases and technologies in order to respond to potential threats in East Asia (and to a smaller degree Central Asia). The geographic constraints presented by the environment of the Pacific Ocean and East Asia are very different from those of Europe and consequently require a different set of technologies, force structure, and strategies. Because of the new map of power and the new theater of action, geography is again at the center of attention.

Reasons for the Decline: Between Natural and Social Science

Geography has disappeared from the study of politics, and international politics in particular, for two reasons. First, the desire to simplify theory led to the abandonment of geography because of its contingency and the difficulty of categorizing it; the growing abstraction of international relations theory led to a devaluation of geography. Second, geographically informed studies have often been marred by determinism that left no space to human actions and freedom. The degenerations that derived from such environmental determinism gave geography, and geopolitics in particular, a bad reputation. However, neither reason for the disappearance of geography reflects reality: geography and geopolitics are unpopular in political science because they have been forgotten, not because they have been conquered by technology and power.[3]

The rediscovery of geography as a powerful explanatory variable has to begin with the study of its intellectual rise and decline. Historically there has been a clear pattern in the study of the relationship between geography and politics, or geopolitics. After starting as a natural science in Europe, with the advent of a theory of international relations in the United States it moved solidly into the social scientific realm. The extreme hypotheses derived from these two approaches submit that geography is either determining (natural science) or irrelevant (social science). Consequently, the passage from natural to social science was a move from geographic determinism to irrelevance.

The reasons for these two hypotheses lie in the different foci of natural and social science. On the one hand, the starting point of a natural scientific approach is the immutable "natural fact" of the geological features of the earth. These facts drive history by forcing human actions in directions predetermined by geographic conditions. In brief, geography as a natural science is a study of the earth and of its influence on human actions. The tendency of a natural scientific approach is to degenerate into a deterministic vision of the influence of geography upon politics.

On the other hand, the tendency of a social scientific approach is to discount the influence of uncontrollable geographic realities and to emphasize the power of human beings. The relationship between men and earth is inverted: it is man who has the power and the ingenuity capable of molding the earth to his wishes. From a social scientific perspective geography is the study of human actions undertaken to overcome the constraints imposed by geography. It is the study of man and his influence on geography.

While the natural scientific approach discounts human agency, the social scientific approach tends to blur the boundaries between the human capacity to conquer geography and the power of geography to influence our actions. At worst, a social scientific perspective on geography fixes no limits upon the creative capacity of men, completely ignoring environmental factors in history and politics.

In the intellectual journey between these two extreme hypotheses one can discern four steps that gradually led to the devaluation of geography as an explanatory variable. First, geographers, or, more generally, theoreticians with a natural scientific perspective, gave the initial push to this study with a decisively deterministic accent: in their view geography determines the distribution of power through climate, resources, natural barriers, and the layout of continents. Second, social scientists, the early Classical Realists, argued that geography was only one among many foundations of national power. Third, a parallel and, to a degree, overlapping school of thought submitted that geography was not a source of, but an obstacle to, power and that consequently it could be conquered. Such argument led to the fourth and final stage, which, claiming that power trumps geography as a determinant of foreign policy and international relations, abandoned entirely the geographic variable.

From Geographers to Social Scientists
THE GEOGRAPHERS

Geographers started the modern study of the influence of geography on politics.[4] Their approach is solidly grounded in a natural scientific method and is characterized by a tendency toward determinism. The point of departure of geographers is the statement that natural forces and the geographic environment have an objective reality, independent of human desires. This reality is largely outside human control, and as such it determines the course of history. Human beings have no choice but to adapt themselves to the geographic characteristics of the environment in which they live.

In the area of international relations, according to this tradition of geopolitical thought, geography determines the distribution of power and productivity by giving a natural advantage to some regions over others. Often the influence of geography boils down to one factor that determines the whole course of history. For instance, climate is perhaps the geographic variable that has been longest, by ancient philosophers and historians such as Aristotle, Pliny, or Montesquieu, as well as by more recent geographers, to explain the differences in the distribution of wealth, productivity, and technological advancement in the world.[5] But other variables—the configuration of seas, the length and direction of navigable rivers, the disposition of continents[6]—have been employed to explain the pattern of history and the shifting locale of power. This intellectual tradition searches for a single geographic cause to explain the course of history and predict the future.

Within this tradition the possibility of changing geography is at best limited. Technological advances can be influential because they modify the value of locations, for instance, by making some places hospitable or by connecting regions that otherwise would be separated. New transportation technologies were especially important in discovering new routes or new continents, altering trade patterns and the location of markets. The acknowledgment of some role for human beings makes geopolitics a dynamic science, as opposed to political geography, which describes static situations.[7] But in the final analysis, for this group of geopolitical writers the geographic environment constitutes a channel of sorts that no human endeavor can overcome and that determines the disposition of power.

The natural scientific approach is characteristic of the early studies of geopolitics. The most well known geopolitical writer, Sir Halford Mackinder, was by training a geographer, and his perspective is heavily influenced by such an approach.[8] In his view, geography determines the fate of nations with a limited input of human actions, though he is careful to avoid "excessive materialism."[9] In the oft-cited book *Democratic Ideals and Reality* he argues that "the groupings of lands and seas, and of fertility and natural pathways, is such as to lend itself to the growth of empires, and in the end of a single world-empire."[10] Mackinder begins by observing that the disposition of continents establishes a central landmass— Eurasia, or the "World Island"—around which all the other continents revolve. Within this landmass there is a geographically inaccessible core, the "Heartland," roughly congruent with today's European Russia. Its state of being landlocked enables the undisturbed development of a strong land power around which those states that control the sea will rotate.[11] Although such a pivotal land power is incapable of reaching the peripheral sea powers—North America, Great Britain, Australia, Japan, and South Africa—it nevertheless presents a constant threat to the shores of the Eurasian landmass. If it succeeds in reaching and controlling an outpost on the Eurasian periphery, it can use its protected land core to generate the power necessary to control the sea and to become the world's hegemon.[12] In brief, Eurasia is an island; if it falls under the control of one power, it will generate enough resources, securely produced in its landlocked geographic core, to master the oceans and dominate the world.

Mackinder goes even further and identifies a geographic point upon which it is possible to base the control of the world. Because the core of the World Island, the Heartland, has a "natural seat of power" in eastern Europe, this small region of Eurasia is the main pivot. The control over this pivotal region is necessary for the control of the Heartland. Hence, the primary geostrategic imperative is to control eastern Europe, the pivot within the pivot. The keys to world domination are here, and not on the southern or eastern periphery of Eurasia. Thus, Mackinder's view can be summarized by his famous adage, "Who rules East Europe commands the Heartland: Who rules the Heartland commands the World-Island: Who rules the World-Island commands the World."[13]

Mackinder's vision is powerful because it derives from geography a single fundamental principle of international politics: the key to world power is the control of the "geographic pivot of history." Geography, not technology or power, determines the locale of this pivot. Technological advances can enhance the capability of the Heartland to project power outside of its core or of the sea powers to conquer the strategic depth of the landlocked "pivot of history."[14] But fundamentally the geographic reality remains unaltered, and the "Heartland" will continue to be the Archimedean spot upon which the history of the world is based.[15]

Mackinder had an enormous influence on the field of geopolitics and international relations. As I will show, Mackinder initiated a study that continued in the United States, although with different accents, with writers such as Nicholas Spykman and Hans Morgenthau. However, because of the strongly deterministic overtones of his theory, Mackinder also became associated with the worst example of geopolitics, the pseudoscience of Nazi *Geopolitik*. The German "science of geopolitics" took the natural scientific approach to a perverted extreme by claiming that geographic features of the earth justified the necessity of Nazi military expansion and human superiority.[16] As one contemporary writer put it, "A large part of Geopolitik is plainly geography gone imperialistic, geography being made to provide arguments and reasons, perhaps even a 'law-like' character, for one's own expansion."[17] Nazi geopoliticians such as Karl Haushofer transformed the study of politics and geography into an ideologically motivated theory. The concept *Lebensraum* was the intellectual product of *Geopolitik*.[18] Because of the dictates of geography, Germany had to expand and seek a "vital space" that was indispensable to her economy and her growing population.

The consequence of the Nazi degeneration was that geopolitics as a viable academic pursuit became a "war casualty."[19] The study of geography and politics was tainted by its association with Nazism, and the entire field was considered to be a perversion from its intellectual inception. As an observer commented in 1942, "In spite of temporary deviations, of slow-downs, of hesitations, the trend of [geopolitical] thought appears to be continuous. . . . [I]t is an unwavering line pointing towards the German goal of supremacy on the continent of Europe, of German expansion towards the great open spaces of Eurasia, with the 'empire of the world in sight.' "[20]

At a minimum, geopolitics was considered to be a betrayal of scientific principles. Because geographers, and among them to a certain degree Mackinder, presented geography as the main, if not only, cause of politics, geography offered an iron law of political action. As another commentator affirmed, "Geopolitics is no more a science than salesmanship is economics. A geopolitician is a geographer who leaves his objective study in order to propose a course of national policy."[21] The connection between geopolitical studies and policy recommendations was dangerous because it not only resulted in extreme forms of imperialism but also ruined a field of science.[22] As Isaiah Bowman writes, "There is no 'science' of geography involved. Geographical facts are marshaled to support political claims. Systems of philosophy are devised which are nothing more than apologia for policies based on military necessity or the logic of 'high culture requires more space.' Map techniques are modified to suit any desired purpose. Clearly this is ideology, not science."[23]

Nevertheless, the Nazi perversion of geopolitics served as a necessary word of caution on the absolute claims of geopolitics, leading to a more sophisticated approach to the study of geography and history.[24] The criticism that the entire field received in the 1940s was in other words helpful because it attracted attention to the tendency of geopolitics to degenerate into extreme forms of environmental determinism. What was needed to correct such a tendency was the recognition of human beings' greater role. As Bowman observed in 1942, "Neither Mackinder nor Haushofer had theories that could stand up to the facts of air power and its relation to industrial strength. Such is the fate of all prophets in this unpredictable world. I might add that the *mind* of man is still a more important source of power than a heartland or a dated theory about it. It is always man that makes his history, however important the environment or the physical resources in setting bounds to the extension of power from any given center at a given time."[25] By stressing the importance of human beings, the criticism generated by *Geopolitik* moved geopolitics away from natural science and closer to social science.

THE EARLY CLASSICAL REALISTS

The gradual passage from the study of geopolitics as a natural science to its study as a social science occurred with the development of international relations theory in the United States.[26] People trained as political scientists, not as geographers, introduced geography into the field of international politics. Two writers, and in particular two books by them, built the foundations of international relations as an academic discipline: Nicholas Spykman's *America's Strategy in World Politics* and Hans Morgenthau's *Politics among Nations*.[27] They started a systematic study of international relations in which geography was constantly present in the background, carefully avoiding charges of environmental determinism by placing greater importance on human action. Theirs was an attempt to take geopolitics out of Nazi hands and restore its reputation.

The first change introduced by these writers was to limit the scope of geography to the realm of foreign policy. Thus, geography was no longer adduced as an explanation for almost every realm of human activity; its role was limited to foreign policy. The geographic features of a state influence its relations with the neighboring powers by making it more or less defensible and more or less apt to expand.

The second change was to consider geography as only one of many variables influencing the conduct of policy and the power of states. In their writings, geography does not have the overwhelming presence of the natural scientific approach of Mackinder or Haushofer. For Spykman and Morgenthau, the early Classical Realists, geography was a form of power alongside natural resources, industrial capabilities, national character, and other intangible qualities. Therefore, although it is "the most stable factor upon which the power of a nation depends," geography must compete with other variables.[28]

In their writings Morgenthau and Spykman distinguish three geographic features: size, borders, and location. First, the larger the territory, the more difficult is to conquer it. Basing his intuition on the historical experience of Russia, Morgenthau observed that "instead of the conqueror swallowing the territory and gaining strength from it, it is rather the territory which swallows the conqueror, sapping his strength."[29] The size of a state offers a passive form of power that makes conquest more difficult "by virtue of being."[30] In fact, in the nuclear age the territorial size of a state has become a key source of power. According to Morgenthau, "In order to make a nuclear threat credible, a nation requires a territory large enough to disperse its industrial and population centers as well as its nuclear installations. . . . [I]t is the quasi-continental size of their territory which allows nations, such as the United States, the Soviet Union, and China, to play the role of major nuclear powers."[31]

Second, the geographic features of the territory of a state influence its ability to conduct foreign policy. For instance, the easier it is to cross a border, the greater the likelihood of an invasion. Morgenthau wrote that the "geographical location of the Unites States remains a fundamental factor of permanent importance which the foreign policies of all nations must take into account, however different its bearing upon political decisions might be today from what it was in other periods of history." The geographic insularity of Great Britain and the United States has created considerable obstacles to intruders, allowing the two island states to develop relatively undisturbed. At the same time, however, borders that are difficult to cross can limit the projection of a state's power, impeding the state's enlargement. For instance, the mountainous borders of Italy and Spain have limited their importance on the European continent by making expansion from their territories arduous but leaving them vulnerable to invasions from the north. "From Hannibal in the Punic Wars to General Clark in the Second World War, this permanent geographical factor has determined political and military strategy."[32]

Third, as Nicholas Spykman observed, "The location of a state . . . will play a large part in determining the political significance of that state, the nature of its international relations, and the problems of its foreign policy." Consequently, "a sound foreign policy must not only be geared to the realities of power politics, it must also be adjusted to the specific position which a state occupies in the world. It is the geographic location of a country and its relation to centers of military power that define its problem of security."[33] States therefore pursue foreign policies on the basis of certain geographic desiderata derived from their location. Furthermore, the geographic position of a state also influences its patterns of foreign policy. For instance, a sea power will extend its influence over a region by conquering a few strategic places necessary to harbor and refuel its vessels, while a land power must expand slowly, almost in concentric circles from its core. "Thus a land power thinks in terms of continuous surfaces surrounding a central point of control, while a sea power thinks in terms of points and connecting lines dominating an immense territory," wrote Spykman.[34]

These early Classical Realists were no longer geographers like Mackinder but not yet pure social scientists. In fact, the passage from the natural scientific treatment of geopolitics to a social scientific one was gradual, and the two approaches coexisted for a while. Harold Sprout observed the moment when these two fundamentally different approaches to geography and politics met. In the early 1960s he wrote that there were two categories of geopolitical hypotheses: "those derived mainly . . . from the layout and configuration of lands and seas, or from regional variations of climate, or from the uneven distribution of minerals and other earth materials" and "those derived from the distribution of people, or from some set of social institutions or other behavioral patterns."[35] One stresses immutable, natural facts; the other, changeable, human factors.

The abovementioned Classical Realists placed themselves in between these two hypotheses. While they looked with sympathy to Napoleon's adage that the policy of every power is in its geography,[36] they introduced several caveats, leaving plenty of ambiguity in the relationship between power and geography. Such ambiguity, I believe, contributed to the abandonment of their intuitive wisdom on the importance of geographic considerations in international relations. In fact, in order to avoid the "land mysticism" of preceding geopoliticians, the independent explanatory power of geography is watered down and becomes indeterminate in terms of both directing interests and bestowing strength.[37] Because geography is only one of the many factors affecting the power of a state, it becomes difficult to discern the influence of geography upon a state's foreign policy from the influence of all the other factors. For instance, the same objective geographic characteristics of the United States allowed American political leaders to pursue a strategy either of isolationism or of expansionism. The insularity of the American continent gave the founding fathers the possibility to watch "the strange spectacle of the struggle for power unfolding in distant Europe, Africa, and Asia."[38] A century and a half later the location of the United States on both the Atlantic and Pacific oceans made American involvement on the European and Asian continents imperative. Obviously, geography alone does not suffice to explain this incongruity in U.S. foreign policy.

There is a similar discrepancy in the power bestowed upon a state by its geography. Depending on which geographic feature is stressed, geography can be a source of strength or weakness. Some Realists, like Morgenthau, argued that the geographic characteristics of Russia (and the Soviet Union) gave her an unmatched advantage over the United States and Western Europe. Russia's size and centrality in Eurasia gave Moscow the capacity to plan a defense in depth, to be within striking distance of the Atlantic and the Pacific, and to spread its industrial production, making it less vulnerable to nuclear attacks.[39] Other Realists, such as Spykman, asserted that geography was not favorable to Moscow because Russia was not located in a controlling position. In fact, it was the "Rimland" (western Europe and Southeast Asia), as opposed to the Heartland, that was a source of world power owing to the superiority of sea power and the location of resources. Paraphrasing the famous dictum of Mackinder, Spykman submitted that "who controls the rimland rules Eurasia; who rules Eurasia controls the destinies of the world."[40]

THE "PURE POWER" REALISTS

Classical Realists tried to navigate between the natural scientific approach, favoring a deterministic view of geography, and the social scientific one, stressing the power of human activity. While Spykman and Morgenthau began to discount the impact of geography, they still maintained an intuitive appreciation for its role in the political life of states. However, both men also sowed seeds of a fundamental change in the approach to geopolitics. Because of its indeterminacy (e.g., size can be a source of power or of weakness, mountains can be a defensive bulwark or a hindrance to expansion, etc.), geography loses its importance in studies on international relations, and the balance of power in particular.[41] When the theorists who followed Spykman and Morgenthau stressed the importance of power as the principal explanatory variable, geography ceased to play a significant role. Indeed, the influence of geography disappeared when power became the main independent variable; geography as an explanatory variable became irrelevant.

The fundamental claim of some Realists was that power, however multifaceted, expands simply by virtue of its nature. The only hindrance to the expansion of power is a countervailing power, understood as material capabilities and leadership skills rather than as geography. States expand when they have power and where there is not enough countervailing power to prevent or arrest the expansion. This is what Arnold Wolfers calls the "pure power model," which is already latent in the thought of Spykman and other early Realists.[42]

The result of this change was that the second, social scientific set of hypotheses described by Sprout prevailed. Human activity and behavioral patterns took over the role played by geography, and geography slowly but relentlessly disappeared as an explanatory variable in international relations theory. Within a "pure power" framework geography is not a source of power or interests but only a constraint that power can surmount through quantitative and qualitative measures. Like a Clausewitzian battlefield, geography offers a certain "friction" that discounts power through distance. Spykman observed that "because the effect of force is in inverse proportion to the distance from its source, widely spaced regions can preserve a certain degree of autonomy but they cannot hope to live in isolation."[43] The *loss-of-strength gradient* can be overcome by sheer quantity of force: if distance x discounts power by a factor y, it is enough to use a quantity of power multiplied by y to make the effect of distance irrelevant.[44] Power trumps geography.

Technological advances can mitigate and finally eliminate the effects of geography. For example, airplanes or missiles can easily cross the most difficult mountain terrain, impassable to infantry, armored divisions, or trade expeditions. The result is that the geographic divisions among states and continents that characterized international relations no longer have an impact on the foreign policy of the great powers, defined as those states that have enough capability to overcome geography. As Spykman observed, "With air power supplementing sea power and mobility again the essence of warfare, no region of the globe is too distant to be without strategic significance, too remote to be neglected in the calculations of power politics."[45]

By becoming a mere obstacle to power rather than a source of it, geography can be conquered and, as a result, lose its importance. It becomes a mere background to the struggle for power between states; it no longer paints the combatants but only the battlefield. This shift in the perspective on geography can be conveyed by a terminological change: geography is no longer the *environment* that influences human actions but a *theater* within which men and states act according to a script written by them.[46]

GEOGRAPHY AS AN "ILLUSION"

The consequence of the shift from environment to theater, from a natural scientific to a social scientific approach, is that the early Realists' intuitive observations on geography, and on distance in particular, are inverted. It is no longer distance that discounts power but power that discounts distance, to the point of making it an irrelevant variable. The belief, in the words of Harold and Margaret Sprout, was that "the trend of human history has been from necessity to conform in order to survive towards ever-enlarging capacity to modify and manipulate environing conditions and events to suit human purposes."[47]

In an influential article published in the late 1960s Albert Wohlstetter argued that distance had no clear discounting effects on power and interests. In fact, the rapid advancement in communications technologies increased the capacity of states to project power to remote areas without a substantial reduction in its effectiveness and amount. Furthermore, these technologies extended state interests to areas that were far removed from a state's borders. The "neighborhood" of states gradually becomes the whole world. Distance becomes an illusion.[48] Within such a framework geography can no longer help predict future actions. Harold and Margaret Sprout wrote about this final stage of geopolitics that "what is being questioned . . . is whether geographic shapes and layouts any longer provide a basis as fruitful as formerly for grand hypotheses intended to explain and to forecast the larger distribution of political potential in the Society of Nations."[49]

The vast majority of current international relations literature is characterized by the absence of geography. Although the perverted versions of geopolitics, notably Nazi *Geopolitik*, are partly to blame for the current dislike of geography, the main cause for the academic irrelevance of geography seems to be the tendency to explain political realities only through political variables. That is, the study of international relations, in particular the study of geography and international relations, has swung from a purely natural scientific to a purely social scientific approach. For such an approach the explanatory variables are social or political, in other words, human variables that exists only insofar as human beings and states exist. Such variables can be anarchy, power, or state identity. Without states (but not without geography) we cannot speak of anarchy or of an international system. Even when this human variable becomes exogenous and uncontrollable by states or men (as in the case of Kenneth Waltz's system), it remains contingent on their existence.

What makes the social scientific approach peculiar is the tendency toward monocausal explanations, similar to that of the natural scientific perspective. The difference lies in the nature or origin of the variable used to explain a political phenomenon. For a natural scientific approach this independent variable is climate, the arrangement of continents, or the location of natural barriers. It is a natural variable, independent of human actions. On the other hand, as mentioned above, for a social scientific approach such single-factor explanations are based on variables such as anarchy or state power. The result is a devaluation, and ultimately irrelevance, of geography.

The difference between the natural scientific and social scientific explanations reflects a fundamentally different assumption concerning the role of human beings. Those who accept the importance of "natural" variables, such as geography, start from the assumption that human beings are inherently limited in their capacity to overcome and make irrelevant the objective reality. Men live and act within an environment that is not amenable to easy changes; they view geography as a hard fact that needs to be taken into consideration. Instead, those who stress the importance of "human" or "social" variables start from the assumption that human beings and the products they create, whether social interactions or technology, are the determinants of state actions. In other words, social science tends to be grounded, as J. David Singer puts it, in "the notion that the only variables exercising any appreciable impact on the behavior of the actors are the internal characteristics, the external behavior, and the social setting of the individuals and groups involved."[50] This is why social science tends to treat international relations as an exclusively social or human realm whose origins and moving factors are to be found in the actors or in the system that they willingly or unwillingly create. "Social" realities are self-sufficient monads that carry within them all the rules and reasons of their functioning.

Kenneth Waltz represents the apogee of this social scientific tendency. In his theory, we do not need to step outside the abstract system of social interactions to understand the behavior of its parts or the nature of the system itself. This is because a theory is a constructed reality that may or may not reflect the actual "outside" reality. As Waltz argues, a theory is an invention constructed in order to explain a law, a causal relationship between two variables. In order to have explanatory power, a theory needs to step back from the observation of reality and create an explanation for that reality. Pure induction leads to an assembly of endless observations and laws but not to their explanation. A theory, therefore, is not a description of reality but a creation of a new reality that simplifies, orders, and explains the plethora of observed phenomena. The farther it is from the actual reality, the more explanatory its power. Hence, a theory constructs "*a* reality, but no one can ever say that it is *the* reality."[51]

It is not surprising, therefore, that in Waltz's theory geography plays no role in influencing the balancing activities of states. The international system is an abstract set of rules or patterns of behavior that forces powers into balancing other powers. The position of a state within the system—not in the real world composed of continents, sea, mountains—determines its actions. This position is established by the amount of power a state has in relation to others, and the geographic location of the various powers does not affect the way the system determines the actions of the states. States are not bound by geographic realities; their only limitation is their own relative power.[52] If the theory incorporates the geographic diversity of the earth, it comes closer to being a description and loses its explanatory power.

Thus, the disappearance of geography from the study of politics has been caused by a drive to make geopolitics, a field spanning natural and social sciences, fully social scientific. This has implied abandoning the examination of the immutable physical facts of geography and concentrating on human interests and actions that are generated independently of the geographic environment.[53] The main lesson is that geography has lost its value with the rise of social science, not with the technological changes in the world. It has been forgotten, not conquered.

Recent Rediscoveries of Geography

Over the past two decades there have been reevaluations of geography in several fields of social science.[54] In international relations theory the need to complicate Waltz's theory and make it more attuned to the reality of international politics seems to be a key motivation behind the revival of geography. If the disappearance of geography is indicative of the rift between theory and policy, the revival of geography is a sign of the desire to bridge that gap. Clearly, theory became so abstract that it lost its value as a roadmap for strategists.[55] Waltz himself conceded that his theory does not predict how individual states will react to systemic pressures.[56]

Nonetheless, while it vanished from theory, geography continued to play an important role in policy-related texts.[57] A geographically informed analysis of international relations offered clear answers to policymakers: where to act, how to project power, where to set up defenses, with whom to seek an alliance. It is in the attempt to reclaim some policy relevance that some theorists are reintroducing geographic variables into the discussion of international relations.

There are at least two schools of thought, represented by Stephen Walt and Robert Jervis, that share an appreciation for geography as a complicating factor in international relations. Both have been categorized as "defensive realism" because they argue that states expand or balance power only when they feel insecure.[58] States do not act simply on the basis of an international hierarchy of relative power but seem to act according to criteria other than a mathematical evaluation of the balance of power.

This approach is an attempt to return to Classical Realists' intuitions, grounded in an appreciation for the multiplicity of reality and in a dislike for simplistic monocausal explanations. It is a return to the realization that states do not act within an abstract anarchical system but within the world. In order to predict and explain states' behavior it is necessary to look not only at what is inside them but also at what is around them. Their location in relation to other states and their geological features are crucial to understanding their behavior. Geography must therefore be taken into consideration as an explanatory variable.

Stephen Walt treats geography in a minimalist way by affirming that distance impacts the perception of threats. Robert Jervis points to the complexity of geography, which affects the likelihood of war by posing obstacles to the projection of military power. Generally speaking, both schools of thought treat geographic circumstances as natural, immutable facts that influence the perceptions or calculations of statesmen. The impact of the environment is therefore limited mostly to the psychological realm. Geography is an intervening variable that affects statesmen's perceptions originating from other sources (power, security concerns, etc.); it mediates the influence of power on the final outcome, whether the foreign policy of a state (Walt) or the international system (Jervis). By doing so, geography mitigates or exacerbates the effects of anarchy.

WALT'S "BALANCE OF THREAT" THEORY: DISTANCE AS A VARIABLE

Stephen Walt uses the basic geographic variable distance, or "proximity," to explain the functioning of the "balance of threat." He begins by observing that power alone is insufficient to explain and to predict the behavior of states. In fact, states balance not against power but against threat, which is a function of four variables: aggregate power, offensive power, aggressive intentions, and geographic proximity.

Walt's argument is grounded in Spykman's (and Boulding's) observation on the inverse relationship between distance and power, the loss-of-strength gradient: power loses its effectiveness with distance. In Walt's balance of threat theory, distance affects the perception of threat: the closer two states are, the more likely they will perceive each other as threatening. "Because the ability to project power declines with distance, states that are nearby pose a greater threat than those that are far away."[59] Because power does not have the same effect everywhere, states do not balance in an abstract realm where a power x has the same importance in place a, b, and c. Instead they balance against their most proximate threat. If states bandwagon with proximate powers, spheres of influence come into being; if they balance, a checkerboard system of alliances arises. Because the predominant response to threats is balancing, neighboring states are more likely to lock themselves in a balancing game than to pursue a bandwagoning strategy.

There are some interesting and policy-relevant consequences of Walt's theory. Given that geographic proximity increases the level of threat, states that are geographically isolated will be less likely to face an antagonistic coalition. For instance, because of its location the Soviet Union presented a greater threat to its neighbors than did the United States, separated from most of the world by two oceans. Consequently, Moscow was forced to deal with a larger number of states that felt threatened and attempted to balance against Soviet power. The result is that "the United States is geographically isolated but politically popular, whereas the Soviet Union is politically isolated as a consequence of its geographic proximity to other states."[60] The policy prescription derived from such a theory is that a geographically isolated state should not attempt to expand in order to avoid spurring an antagonistic coalition. Geographic isolation is a strategic blessing and should not be squandered by an expansionary strategy.

Walt's theory is an important refinement of neorealism. By reintroducing the concept of a loss-of-strength gradient, he brings the international system closer to reality. Nonetheless, this perspective on geography and on its influence on the foreign policy of states has two limitations. First, Walt accepts only one geographic characteristic, distance.[61] It is probably the easiest geographic variable to categorize and quantify, and as such it is very appealing. However, the influence of distance alone upon threat perceptions is unsupported by strong empirical evidence. For instance, in the mid-1930s Paris and London did not share similar perceptions of a growing German power even though they are roughly equidistant from Berlin.[62] A reason for this difference in threat perception was the difference in the "nature" of the distance between these states. The existence of the Channel offset the growth of German power for the British, while the lack of powerful natural barriers (further weakened by the German occupation of the

Rhineland) encouraged the French to forge alliances with Poland and Czechoslovakia. Distance alone, therefore, is a weak geographic variable.

Second, Walt considers only distance between homelands. Although territorial security indubitably is the preeminent concern of states, it is not the only motivation spurring balancing coalitions. Because states are not self-sufficient, their security often depends on the supply of raw materials or other resources from distant countries. A threat to those sources of power, no matter how removed from a state's homeland, represents a direct threat to the state itself. In other words, the geographic extent of a state is not limited to its home territory but encompasses areas vital to its survival. For instance, for the United States, the Persian Gulf is one such area because of the importance of oil to the American economy and military machine. A threat to this region, coming from the Soviet Union or Russia or from internal rivalries, posed, and continues to pose, a direct danger to the United States. Thus, even if we use distance as the geographic variable, the territory of a state cannot be the only starting point.

OFFENSE-DEFENSE: GEOGRAPHY AS A MEDIATING VARIABLE

Robert Jervis and the literature that he spurred take a slightly different approach to geography. First, their geographic variable is more complex than mere distance because it includes geological features (mountains, oceans, rivers, etc.). Second, the influence of this multifaceted geographic variable is systemic, affecting the balance of power and the likelihood of war rather than alliance preferences and foreign policies of states.

In his seminal article "Cooperation under the Security Dilemma" Jervis argues that geography plays an important role in exacerbating or mitigating the security dilemma. The security dilemma is faced by all states pursuing their own security. By trying to achieve greater security, states act in ways that are perceived by others as threatening and as decreasing their own security. Geography is a mitigating variable because, together with other variables, it can lower the feeling of insecurity of all actors, diminishing the need to undertake retaliatory military actions. For instance, a state that is geographically isolated and hence relatively secure will not need to pay much attention to political developments of other nations because sheer geography makes them less threatening. Consequently, the security dilemma of such a state (e.g., Great Britain or the United States) is less "acute" than that of one situated in a geographically vulnerable position (e.g., the Austro-Hungarian Empire).[63]

Geography affects the security dilemma by influencing the offense-defense balance. In the "offense-defense" literature that sprung from Jervis's article geography plays a mediating role between states' actions and systemic outcomes by increasing offense or defense dominance in international relations. When there are sufficient natural defensive barriers (e.g., oceans, large rivers, and high mountains), the defense has the advantage, and the security dilemma is less intense because the threat posed by states pursuing their own security is smaller. When such barriers are missing (e.g., on the flat central European plains), the security dilemma becomes acute, with a resulting high risk of war. Thus, natural barriers, large territory, possession of strategic resources, and security of trade routes endow a state with great defensive capability. Conquest becomes more difficult, and anarchy is less volatile.[64]

One of the principal findings derived from this insight is that anarchy is not equally dangerous in every corner of the world. Some regions, like Central Europe, are more likely to be rocked by chronic insecurity because of the geographic proximity of states and the lack of geographic barriers to offensive actions. Other locations, like the British Isles, are endowed with a defensive geography that lessens the security dilemma. As Jervis observed, "If all states were self-sufficient islands, anarchy would be much less of a problem."[65]

This is an insightful and welcome revival of geography. It has, however, two weaknesses. First, in many cases geography is constant but the offense-defense balance shifts. Hence, the main source of change must reside in other variables, like military technology, which determine the final outcome of the offense-defense balance. The equilibrium between offense and defense is based on "castles versus cannons, machine guns versus trenches."[66]

Second, the offense-defense literature, like Walt's theory, limits the geographic extent of the state to its homeland. This allows it to argue that if the state is endowed with defensive geographic features (e.g., an island), it functions within a less anarchical environment. Nonetheless, such a state might be forced to pursue a very aggressive foreign policy because it needs resources that are not found in its home territory. There is no perfect "island": all states, to different degrees, must seek strategic resources outside their territory. In this way they effectively extend their territory, losing the stability and security derived from their isolated geographic position.

From this brief overview of some recent attempts to "bring geography back in" it is evident that the renaissance of geography in international relations is still in its early stages. The main weakness of this literature, and of the geopolitical tradition, is the lack of a clear definition of *geography*. The minimalist definition considers only distance; the maximalist takes into account every geographic feature. Both are problematic: distance is easy to quantify but difficult to use as a causal variable (not to mention that distance is probably the easiest geographic variable to conquer through technology), while geographic features are impossible to generalize because every region in the world is unique. The next task, therefore, is to clarify what I mean by *geography*.

CHAPTER TWO

Geography, Geopolitics, and Geostrategy

Any study of the influence of geography on politics must begin with the definition of the former. But to define *geography* is both difficult and dangerous. The difficulty lies in the fact that if geography must have explanatory power, it has to allow a degree of variation: a constant reality, such as the geological features of the world, cannot explain the often dramatic changes in foreign policy or in the political fate of states. The danger lies in the fact that unclear definitions of *geography* make it either deterministic or irrelevant.

It is imperative, therefore, to clarify the concepts used and identify the geographic variables. My argument revolves around three concepts: geography, geopolitics, and geostrategy. These concepts are determined by three geographic variables: trade routes, centers of resources, and state borders. The first two affect geopolitics, and the third influences geostrategy.

Three Concepts: Geography, Geopolitics, Geostrategy

States must reflect the underlying geopolitics in their foreign policy or geostrategy. When they fail to do so, the state's political success and even political survival are at risk. Only states that pursue a geostrategy reflective of geopolitics gain and maintain an advantage in their relative power. My argument, therefore, hinges on the relationship between geography, geopolitics, and geostrategy.

Geography is the physical reality, composed of mountains, rivers, seas, wind patterns, and so on. It describes the geological features of the earth, the physical attributes of the land, sea, and air environments.[1] With a few exceptions, such as natural disasters (seismic activities, climatic changes, etc.), which change the geological features of a place,[2] or dramatic political changes (imperial expansion, boundary adjustments), which alter the geographic setting of a state,[3] geography is a constant. Consequently, by itself it is not a useful variable to explain variation in foreign policy.

Geopolitics is the human factor within geography.[4] It is the geographic distribution of centers of resources and lines of communication, assigning value to locations according to their strategic importance.[5] The geopolitical situation is the result of the interaction of technology broadly defined and geography, which alters the economic, political, and strategic importance of locations. For instance, new routes are discovered or, literally, carved out in mountains thanks to the development and implementation of new communications technologies. Similarly, differentials in economic growth alter the distribution of power in the world, while the introduction of new production technologies changes the need for natural resources. Geopolitics therefore is not a constant but a variable that describes the changing geographic distribution of routes and of economic and natural resources.

Geostrategy is the geographic direction of a state's foreign policy. More precisely, geostrategy describes where a state concentrates its efforts by projecting military power and directing diplomatic activity. The underlying assumption is that states have limited resources and are unable, even if they are willing, to conduct a *tous azimuths* foreign policy. Instead they must focus politically and militarily on specific areas of the world. Geostrategy describes this foreign-policy thrust of a state and does not deal with motivations or decision-making processes. The geostrategy of a state, therefore, is not necessarily motivated by geographic or geopolitical factors. A state may project power to a location because of ideological reasons, interest groups, or simply the whim of its leader.

One way to conceptualize geography, geopolitics, and geostrategy is by examining their patterns of change. There are three different levels of change, ranging from tectonic (no change) in the case of geography to potentially rapid change in the case of geostrategy. Geographic changes are measured in geological ages of thousands of years, while geostrategic changes are measured in days, months, and years. As mentioned above, geography is by and large constant, with the exception of catastrophic events that are rare and unpredictable. Geopolitics changes with the rise and decline of centers of resources and shifts in routes. It is a change that occurs slowly, often imperceptibly, and usually spans decades and centuries. The late-fifteenth-century discoveries of new routes around Africa, linking Atlantic Europe directly with Asia, are an example of a geopolitical change that over the course of a few decades altered the map of the world. The current economic growth of East Asia, and China in particular, in a few years may represent a geopolitical change of similar proportions.[6]

Finally, geostrategy is the most flexible of the three concepts. It can change quickly, in weeks or months, following bureaucratic processes or changes in

TABLE I
Geography, Geopolitics, and Geostrategy

		Change	
	Level	Type and Cause	Effect
Geography		Tectonic—de facto constant	
Geopolitics	Systemic	Slow—rise and decline of empires; new transportation and production technologies	Changes in strategic value of locations, trade routes
Geostrategy	State	Varied—dependent on situation on state borders	Success—reflective of geopolitics; failure—nonreflective of geopolitics

leadership. For example, the decision of Ming rulers in the mid-fifteenth century to end maritime expeditions to the Indian Ocean and East Africa represents such a dramatic and sudden geostrategic reorientation. Similarly, the U.S. invasion of Afghanistan in 2001 is another example of a dramatic change in the geographic focus of U.S. foreign policy; a theater that for decades was considered irrelevant by the United States suddenly became the focus of attention.

Geography, geopolitics, and geostrategy constitute in a sense three layers of the international arena that move at different speeds and for different reasons. They are related to, but do not determine, one another. Geostrategy is not a mere reflection of the underlying geopolitics, which in turn is not a copy of geography. Conversely, geography does not determine the geopolitical situation, which in turn does not determine the geostrategies of states. Geopolitics describes the geographic distribution of centers of resources and routes, which, however, is determined by a combination of technology and geography, and not by geography alone. Similarly, geostrategy is an interpretation and a response to geopolitics and is not determined by it.

Because of the different patterns and sources of changes, geography, geopolitics, and geostrategy are not always "aligned." The study of this "alignment," and of its importance for the relative power of states, is the crux of my project. In particular, I am interested in the relationship between geopolitics and geostrategy: How can a geostrategy reflect the underlying geopolitics? What are the consequences when it fails to do so?

The challenge for strategists is that geostrategy does not automatically reflect geopolitics. This is made evident especially in moments of great geopolitical changes, when those changes are not followed by appropriate changes in geostrategy. That is, geostrategy often does not adjust to geopolitical change, either because of a leadership failure or because of the location of the state in question.

Statesmen may fail to read geopolitics and geopolitical change correctly and thus do not formulate and implement an appropriate response to changes. The decision of Venetian strategists to expand on the Italian *terraferma* in the fifteenth century and the withdrawal from the sea ordered by Ming emperors roughly in the same period are examples of such failure.

Furthermore, strategists are often hindered by the geographic position of the state. The location of a state, for instance, makes it difficult, if not impossible, to devise a geostrategy that reflects a dramatic geopolitical change. The sixteenth-century reorientation of trade routes following the discovery of America and the Cape of Good Hope passage made the Mediterranean, and Venice in particular, strategically less pivotal. Arguably, there was probably little that Venice could do to change such a situation. Although I do not argue that geography determines the fate of states, it certainly limits their strategic choices and their ability to adapt to a new distribution of power in the world.

When there is a disconnect between the geostrategy of a state and the underlying geopolitics, that state begins its decline. The state loses control over centers of resources and lines of communication and consequently relinquishes much of its influence over other states. This is not unavoidable, for states can and do change their foreign policy to reflect more adequately the geopolitical situation. But as illustrated in the case studies, diplomatic, technological, and bureaucratic challenges often make it difficult to reorient geostrategies.

Geopolitics

Simply stated, geopolitics is the world faced by each state. It is what is "outside" the state, the environment within which, and in response to which, the state must act. More precisely, geopolitics, or the geopolitical reality, is defined by lines of communication and by the disposition of centers of economic and natural resources. These two variables, in turn determined by the interaction of geological features and human actions, create a set of objective and geographically specific constraints to the foreign policy of states. In brief, geopolitics is an objective reality, independent of state wishes and interests, that is determined by routes and centers of resources.

The Objectivity of Geopolitics

The first characteristic of geopolitics is its objectivity. By this I mean that geopolitics, or the geopolitical situation, exists independently of the motivations and power of states and is not contingent on the perceptions of strategists and politicians.

States cannot alter geopolitics to match their interests, or at best they are very limited in their capacity to do so. A change in geopolitics involves a change in routes or in the location of resources, and a state cannot single-handedly effect such a change. Geopolitical shifts follow changes in production and transportation technology, which occur over the course of decades and are rarely controlled by a single country.

Geopolitics is distinct from and independent of the perceptions of strategists. Such perceptions, and the resulting decisions and actions, do not shape the geopolitical reality but only respond to it. They are interpretations of it. As Harold and Margaret Sprout observed in a paper on the relationship between man and milieu, "The limitations, like opportunities, are latent in the milieu, but inoperative until some decision is taken and implemented. These limits vary from place to place, and from one historical period to another. But limits there indubitably are, limits which will affect the outcome of any course of action undertaken, irrespective of whether or how perceived and reacted to by the actor in question."[7]

It is very difficult to "discover" geopolitics and consequently to formulate the most appropriate geostrategy. The difficulties are twofold. First, there is no perfect geopolitician, someone who can see the geopolitical situation without any other interests or ideas clouding his vision. All these factors mediate between geopolitics and geostrategy.[8]

This is not to say that the perceptions of individual strategists are irrelevant. In fact, such perceptions shape the foreign policy of a state, directing its attention and power toward areas that are deemed strategically important. As Nicholas Spykman observes, "Every Foreign Office, whatever may be the atlas it uses, operates mentally with a different map of the world."[9] The challenge is that when these mental maps, vitiated by ideological principles, domestic political concerns, or mere incompetence, do not mirror the underlying reality, they lead to a bad foreign policy, projecting power to areas that are perceived as important but are in reality strategically irrelevant. In other words, these subjective maps shape the foreign policy of a state but not the geopolitical reality faced by it.

The second difficulty of reflecting geopolitics in the formulation of geostrategy lies in the nature of geopolitical change. Unlike geography, geopolitics is in constant flux, with some routes becoming more important than others, while old centers of resources are being replaced by new ones.[10] These changes occur at tectonic speeds and are better categorized as long-term trends rather than one-time catastrophic events. As a result, they are extremely difficult to gauge while happening and become clear only after they have occurred.[11] Strategists might be able to determine the principal lines of communication and the regions richest in strategic resources at any given point in time, but concerning the direction of geopolitical changes they are forced to rely at best on forecasts derived from history. For instance, the technological switch from a coal to a diesel engine not only did not happen overnight but also did not bring an immediately corresponding change in the strategic importance of regions rich in coal and oil. It took decades before the Caucasus, the Caspian Sea, and the Middle East supplanted Wales and Silesia, and it can be argued that this switch has never been complete because coal continues to be an important source of energy. The passage from one period to another, from one geopolitical situation to another, is not marked by well-defined moments that unavoidably alter the world; it is often difficult to establish when one ends and the other begins. These periods are visible in a broad historical sweep or in a study of "macrohistory."[12]

Two Variables

If it is so difficult to "read" geopolitics, then what is the relevance of studying it? Or, more practically, how can one discover the underlying geopolitics and trace its change? If historians can identify specific periods, for instance, the "Vasco da Gama" era, which led to the rise of the Atlantic and the decline of the Mediterranean, then can a strategist do the same and act upon it? I argue that in order to discover the geopolitical reality it is necessary to look at the location of resources (distribution of power) and the lines of communication linking them.[13] The configuration of these two variables assigns strategic value to locations, privileging some over others.

LINES OF COMMUNICATION

Lines of communication or routes link states with one another.[14] Relations between states consist of commercial exchanges, military clashes, and information exchanges, all of which flow through well-defined channels determined by geography and technology. Lines of communication in a sense constitute the nerve system of the world, through which international relations occur. Most international exchanges, from the oil trade to information flows, are linked to geography. For example, one-third of total U.S. trade by value with the G-7 countries is sea based. Moreover, global maritime trade is growing at a steady annual rate of 3–4 percent, making sea lanes increasingly important.[15]

Furthermore, lines of communication are important because through them states project power and access centers of resources. The bulk of military power is still projected via land or sea, while over the past century the amount of logistical support needed by armies has increased exponentially, increasing further the strategic value of routes.[16] In 1914 an infantry division used about one hundred tons of supplies a day; in 1940 a German armored division needed three hundred tons a day, and just a few years later a similar American division consumed twice as much. Currently estimates range from one thousand to fifteen hundred tons of supplies per day.[17] These supplies are shipped via sea, land, and air routes, underscoring the importance of safe lines of communication.[18]

A state that controls lines of communication has full strategic independence. It does not have to rely on the goodwill and protection of other states to access the resources it needs, project power where it wants, and maintain commercial relations with whom it wants. When a state does not have control over the routes linking it with the source of resources and other strategic locations, it falls under the influence of the power in charge of those lines of communication. This is why control of routes has always been an objective of states.[19]

An illustrative example of the weakness caused by the lack of control over routes is the Russo-Japanese War, in 1904–5. At the time Russia was locked in a struggle with Japan, which by May 1904 had succeeded in encircling the Russia's farthest Asian outpost, in Port Arthur. In an attempt to lift the siege, Russia hastened reinforcements across her territory, but because the Trans-Siberian Railroad was not yet finished, troops could not be sent in sufficient quantities to repel the Japanese onslaught. Hence the only way of relieving Port Arthur was by sea. Thus Russia sent her Baltic Fleet to the Pacific theater to help the beleaguered army and navy in Korea.[20] The twenty-thousand-mile voyage required several coaling and resupplying stops for the ships, and Russia did not control the sea lanes around Africa and was forced to rely on German and French colliers and colonial ports. The Baltic Fleet, therefore, was at the logistical mercy of the German Hamburg-America Line colliers (coaling the Russian ships at neutral ports but outside of the territorial waters) and of the French bases in Africa (Dakar, Gabon, Libreville, Madagascar) and Indochina (Camranh).[21] While the technically and tactically inferior Russian navy was soundly defeated by the Japanese, the need to use the bases of other powers put Russia at their mercy. Germany and France used this opportunity to pursue their own diplomatic games, aimed mainly at increasing the Anglo-Russian enmity, already made volatile by the Russian threat to India and by an Anglo-Japanese alliance, and at creating some form of Continental entente with the British power.[22] As observers commented after that war, Russia's geography and lack of an adequate network of bases prevented her from being a great power.[23]

It is important to clarify that sea lanes are not the only routes that are vital to states. Historically, especially since the sixteenth century, sea lanes have been the principal channels of the world economy because of the relatively small costs of long-haul shipping. Because of the cost-effectiveness of sea shipping, A. T. Mahan could assert that sea powers, those states that controlled sea lanes, were naturally superior to land ones.[24] Mahan's point cannot be taken to the extreme, however, because despite the importance of maritime routes, there were and continue to be vital lines of communication on land. The Silk Road, which during the Middle Ages linked Europe with East Asia through Central Asia, is a good historical example of the importance of a continental route. More recently, oil pipelines, railroads, and highways represent important continental routes and must be taken into consideration when examining the geopolitical reality.

Moreover, focusing on sea lanes tends to minimize the importance of controlling lines of communication in general. Most sea lanes are not under the sovereign control of a state, and as a result it has often been assumed that they were free and equally accessible to all states, especially during periods characterized by the maritime hegemony of one power (e.g., Great Britain or the United States). However, maritime powers have the ability to control the sea lanes and to deny or threaten the free passage of ships of other powers. The fact that sea powers tend to exercise such capability of denial only in cases of conflict leaves the illusion of a free maritime space. The current global situation is a good example of a maritime hegemon, the United States, guaranteeing the free flow of goods on the seas. Like past maritime powers, such as Venice or Great Britain, the United States can deny other states access to key sea lanes as well as safeguard a sea lane in case it is threatened by political instability or an enemy.[25] Freedom of passage does not mean absence of control.

Finally, the configuration of routes changes on the basis of three variables: the discovery and creation of new routes, changes in transportation technology, and changes in the location of resources. These three variables are analytically distinct but are also intertwined because, for instance, new transportation technologies make possible the discovery of new routes, which in turn can alter the distribution of power in the world. The main point is that routes are determined not only by geography but also by human actions; consequently they are not constant and have shifted throughout history.[26]

The discovery and creation of new routes lead to the most spectacular and quick changes in the configuration of routes. Probably the most momentous shift of routes, one that continues to define the modern era, was the late-fifteenth-century discovery of the Atlantic sea lanes connecting western Europe with America and, by circumnavigating Africa, with Asia. Such discovery undermined the importance of land routes crossing Central Asia because it offered a cheaper and, with the rapid technological improvements in shipping, more reliable means of transport.

Although the age of geographic discoveries is over, new routes continue to be established. For instance, the excavation of canals, such as the Panama and Suez canals, can unite bodies of water that otherwise would be separated, altering their importance.[27] Similarly, the construction of tunnels or internal waterways extends the transportation system to new locations by shortening shipping routes and decreasing their costs.

Routes can also change through the invention and implementation of new transportation technologies. New technologies make the utilization of new routes possible.[28] The invention of the airplane, for instance, heralded the advent of a new medium of transport, establishing previously unknown air routes. But such inventions are akin to geographic discoveries in their momentous implications as well as their rarity. Most often, technological changes affect the configuration of existing routes. For example, the switch from coastal, oar-rigged galleys to ocean-worthy sailing ships in the fifteenth and sixteenth centuries and, in particular, to coal-fueled vessels in the nineteenth century, with corresponding advances in navigation instruments and skills, changed the logistical requirements of fleets, altering the configuration of maritime routes. Ships no longer had to hug the coast but could hop from port to port along the route, shortening the navigation time and costs. Longer routes, such as those circumnavigating Africa, became feasible and cost effective.[29]

A similar change in the configuration of routes occurred with the development of railroads in the nineteenth century. Before the advent of railroads in Europe it was cheaper to send goods from northwestern to southern Germany "by ship from the northern ports, through the English Channel, around Gibraltar, through the Dardanelles, and up the Danube to their destination."[30] The undeveloped Continental transportation network gave a strategic advantage to Great Britain, which controlled the sea lanes around Europe and could blockade Continental economic exchanges. In the second half of the nineteenth century the growing rail system shifted internal European trade, in particular German trade, from the sea to the Continent, curtailing British influence over Berlin's domestic market.[31]

Finally, routes are strategically important insofar as they grant access to resource-rich regions.[32] When a resource-rich area, whether a coal-mining region or an industrial zone, declines in importance, the routes connecting it with the markets and other states become obsolete because there is no need to ship goods from the region or to project power to it. Similarly, the rise of a new resource-rich area leads to the search for new routes. For instance, the collapse of the Soviet Union opened the Caspian Sea region to investment in new oil fields, increasing its oil-producing potential. The post–cold war rediscovery of the strategic importance of this area led to the search for new routes that could guarantee access to it. For the past decade several states, including Russia, the United States, and Turkey, have been engaging in a new "great game," vying for the construction of pipelines that would give them control over access to the Caspian natural resources.[33] Lines of communication therefore must be adapted to link new centers of resources with the rest of the world because routes that connect two irrelevant regions are themselves irrelevant.

CENTERS OF RESOURCES

The importance of resources for the state has been noted by innumerable theorists.[34] Resources are the best proxy for power because not only are they easy to quantify but their abundance correlates with a powerful state. Natural and economic resources fuel a state's industrial and military capacity and consequently are strategic goods the control over which bestows influence and power.[35] Moreover, because they are distributed unevenly, resources make some regions strategically more valuable than others. For instance, thanks to their economic and natural resources, Hong Kong and Kuwait are geopolitically more relevant than Crete or Mongolia. What happens in and to such regions has an impact on the lives of other states, which consequently will pay more attention, militarily or diplomatically, to them.

Niccolò Machiavelli succinctly summed up the importance of resources when he wrote that "it is necessary in the founding of a city to avoid a sterile country. On the contrary, a city should be placed in a region where the fertility of the soil affords the means of becoming great, and of acquiring strength to repel all who might attempt to attack it, or oppose the development of its power."[36] The intuition behind Machiavelli's statement (shared by other political theorists, from Plato to Morgenthau) is that a state is secure and powerful when it can achieve autarky. As one of the fathers of Realism, Hans Morgenthau, put it, "A country that is self-sufficient, or nearly self-sufficient, has a great advantage over a nation that is not" because it does not depend on the will or power of other states. As an illustration he mentions that during both world wars one of Germany's main

goals was to destroy British sea power, which threatened to close sea lanes supplying Berlin with foodstuffs.[37] It is important not only to own resources but also to
have free access to them.

There is a growing debate on the political and economic benefits of exercising
direct control over resources, in particular economic ones. While there is no disagreement on the strategic importance of both natural and economic resources,
many argue that it is not necessary, and perhaps even not possible, to control
directly such resources in order to extract political, economic, and military benefits. The argument is grounded in the belief that an increasingly globalized economy allows states to buy the necessary goods, and consequently free trade decreases the need and the incentive to extend direct political control over them.
Moreover, especially with regard to modern industrial capacity, direct (or imperial) control over economic resources does not result in the accumulation of
wealth and power. It simply might be impossible to accumulate industrial capacity through conquest because of certain characteristics, such as the geographic
dispersion of production and the knowledge-based sector, of the modern global
economy.[38] Even some Realists argue that military conquest in the modern era no
longer brings an accumulation of power.[39] As Robert Gilpin writes, "In the modern era, expansion by means of the world market economy and extension of
political influence have largely displaced empire and territorial expansion as a
means of acquiring wealth."[40] The goal of modern empires, therefore, is to establish and enforce rules, which in turn create regional or global markets through
which resources are obtained.[41]

Although important, asking whether conquest is profitable misses the point.
The conquest of resources may or may not pay, but the denial of access to them is
still a source of power and is still sought by states. That is, a state might be unable
to obtain sizeable benefits from owning a resource-rich territory, but by simply
preventing other states' access to it, it can exercise enormous political leverage. As
Klaus Knorr observes, "Economic power is used coercively by threatening to deny
some sort of economic advantage to another state, often but by no means necessarily for the purpose of gaining an economic benefit." He adds that "the ability to
shut off valuable markets, to preempt sources of supply, to stop investments, or to
reduce economic aid constitutes bases of national economic strength."[42] The
possession of wealth, broadly defined, or the denial of access to it is a source of
power when it is used to deliberately change the behavior of others. For example,
the U.S. policy of containment, especially in George Kennan's formulation, was
based on the premise that resource-rich regions were strategically vital. Kennan
argued that if the United States protected the vital power centers from Soviet

aggression, it could not only contain but also defeat Moscow.[43] The United States did not have to conquer these centers; it was sufficient to deny them to the Soviet Union. Denial of access to resources can be more powerful than their conquest.

There are two types of resources: natural and economic. Natural resources are goods like timber, coal, oil, or water, to mention those that historically have been the most important. They are the geological wealth that comes with the territory. Economic resources are industrial goods, such as steel, machines, and other manufactured products. They are the wealth that is created by people. Economic resources are not necessarily associated with natural resources. A state with limited natural resources can be an industrial power (e.g., Hong Kong), just as a state rich in natural resources can remain economically underdeveloped (e.g., Iran or Russia).

It is important to distinguish between these two types of resources because they change in different ways. How they change affects how, and how fast, geopolitics changes. The rise and decline in importance of a specific resource, such as oil or coal, alters the value of the regions where that resource is located. For instance, the Persian Gulf region was not always a vital source of natural resources, while Europe's Atlantic states became economically important only in the fourteenth and fifteenth centuries. Changes in the importance of resources have the double effect of altering the strategic value of locations and consequently of the routes connecting them.[44] A route linking locations rich in resources that are not in high demand is strategically irrelevant.

In the case of natural resources, it is relatively easy to explain changes in their geographic distribution but not so easy to predict those changes. They occur because of either discoveries or technological changes. Unlike routes, which by the beginning of the twentieth century had all been mapped, the discovery and exploitation of natural resources continues to be a key source of geopolitical change. The discovery of a new oil field, for instance, can quickly transform an irrelevant region at the margins of geopolitics into an area of utmost strategic importance. The ongoing debate over sovereignty over the Spratly Islands, in the South China Sea, which are reported to have oil, is an example of a geopolitical change that would alter the value of a region.

Improvements in production technologies can also change the geographic distribution of natural resources. Technological inventions, such as the diesel engine, create the demand for new resources and consequently increase interest in the regions rich in those resources. As the economists Eric Jones, Lionel Frost, and Colin White aptly put it, "There are no resources as such, only the possibilities of resources provided by nature in the context of the technology of a given society at a certain moment in its evolution."[45] Strategic locations lose their importance because new technologies require new natural resources (e.g., coal, oil, or water).[46]

An example of the impact of technological advances on geopolitics is the introduction of the diesel engine in the nineteenth century. The result was a shift from coal to oil as the principal source of energy. This shift was gradual; only in the second decade of the twentieth century did the adoption of oil-fueled ships make oil an indispensable strategic resource.[47] By the end of World War I, oil was required to run the military machines of the belligerents.[48] The geopolitical shift was gradual because coal continued to play an important role in providing energy. It was only after World War II that oil became the strategic resource par excellence. In the United States oil consumption increased gradually until it surpassed the use of coal in 1950.[49] Great Britain was a bit slower in switching to oil because of heavy government subsidies to the coal industry and a low-wage mining labor force, but in 1971 oil surpassed coal as the main source of energy for the British economy.[50] As the global consumption of oil steadily surpassed that of coal, new regions of the world became important. The Caspian Sea region at the turn of the century and later the Middle East became strategically more important than the coal-rich Ruhr, Silesia, and Wales.[51]

The rise and decline of centers of economic resources is a complex subject that encompasses a vast area of human activity and by necessity spans centuries and introduces multiple causes. To explain these macrotrends in the geographic shifts of power it is sometimes necessary to simplify the causation in history, often reducing it to often a highly abstract concept such as "culture" (e.g., Max Weber's Protestant ethic) or "geography" (e.g., Mackinder or, even more recently, Jared Diamond).[52] Changes in economic centers are visible only in what the historian Fernand Braudel termed *la longue durée*, that is, a timespan of centuries. As a result, the questions asked are very broad: Why do some regions become more prosperous than others? Why, for instance, did the Atlantic rim of Europe attain higher economic growth than Central Europe starting in the fifteenth and sixteenth centuries? Or, a question that has stirred several debates recently, why did Europe develop faster than Asia?[53] These questions try to determine the reasons behind a particular geographic distribution of economic power.

The answers to these questions are diverse. They focus either on the conditions that favored economic growth or on the obstacles that hindered it. Broadly speaking, there are four categories of explanations: geographic, cultural, state-centric or institutional, and, for lack of a better term, fatalistic.

The economic growth (or decline) of a region or state is associated with the

presence of certain geographic conditions. For instance, the possession and exploitation of natural resources facilitates growth. The United States, for example,
built its industrial base largely on its superior geological endowment.[54] Conversely, tropical climates accompanied by highly infectious diseases and landlocked locations hinder economic development.[55]

A nation's or region's culture, broadly and often vaguely defined, can be more
or less conducive to economic development. For instance, following Weber's
argument, some have found the Protestant ethic a source of western European
economic growth since the sixteenth century. Along similar lines, some argue
that Confucian values, which privilege obedience to a rigid social hierarchy and
isolation from the outside world, stifled innovation and commerce, eventually
leading to the economic decline of China.[56]

The role of the state, in particular the interaction between state institutions
and the market, is the principal factor behind the economic growth of states. A
vibrant market develops when states create institutions, or more generally conditions, that allow a middle class to prosper through innovation, production, and
commerce. However, when states are overbearing, forcing societies to focus on
the interests of the political center, they are unlikely to be economically competi-
tive in the long run.[57]

"Fatalistic" explanations constitute a world-view that tinges the abovementioned explanations more than explanations in and of themselves. They are characterized by a structural fatalism, often leading to a cyclical view of change. A
variety of domestic and international structural variables, independent of the
actions of individuals, determine the rise and decline of power. On the international level, diffusion of technology, overextension of the state or empire, and
the imperial burden of defense are some of the variables adduced to explain
growth or decline.[58] On the domestic level, the internal structure of resource
extraction and the moral fabric of society influence the power of the state. These
variables carry a certain inevitability that forces states, in particular those that are
economically powerful, into a cycle of rise and decline. As a result of such cycles,
the geographic locus of power is constantly and inevitably changing.

These four broad categories explaining geographic changes of economic resources are by no means exhaustive. The rise and decline of states and the geographic migration of power have generated prolific studies that often do not fit
into one of the above categories. The complexity of the change in the distribution
of economic resources is probably best explained by a multicausal argument that
can indicate the conditions that favored or hindered economic growth.

It is not my goal, however, to debate the merits of these explanations or to offer a more comprehensive account of the rise and decline of economic powers. I limit my argument to an acknowledgment of the fact that economic resources shift geographically and that it is very difficult to predict the geographic direction of such shifts. For whatever reason, whether the decline of the national character or the costs of the defense apparatus, the location of the current economic power might be different from that of the past or future. For instance, the core of the world's economic power was located in the Mediterranean region until the fifteenth or sixteenth century, when it shifted toward western or Atlantic Europe. Similarly, over the past half-century East Asia has become an important center of economic power, dramatically altering the distribution of wealth and power in the world. As Jones, Frost, and White observed, "The rise of the Asia Pacific economy is the dominating fact of modern economic geography, the most striking event in the economic history of the late twentieth geography. On the face of things, this is a virtual break in the trends of history—the first achievement of sustained growth by a major cultural area outside Europe or regions of European settlement."[59]

Thus, the geopolitical situation is defined by lines of communication and centers of natural and economic resources. Their geographic distribution determines the strategic importance of regions, the control of which bestows economic power and political leverage. Geopolitics is the map within which states, whether they perceive it correctly or not, act.

While geopolitics is defined by two variables, trade routes and the location of resources, only one of them is necessary to change the geopolitical situation. For instance, a shift in trade routes without a corresponding change in the geographic configuration of resources is enough to create a new geopolitical situation. The state that controlled the old routes loses leverage because of competing lines of communication. But as I pointed out earlier, often such changes are interconnected because, for example, a change in resources also alters trade routes, and vice versa.

There is no clear trend in geopolitical changes. Geopolitically inclined theorists often fall into the trap of wanting to find, and claiming to have found, a discernible direction of geopolitical change. Some have argued that the strategic advantage has been shifting historically from land to sea power. (This is Mahan's overarching argument.) Others have claimed that the locus of power has moved inexorably from east to west or from the heartland to the rimland of Eurasia. I do not see any linear progression of geopolitical change. The development of new communications technologies does not necessarily lead to a greater advantage of sea powers over land powers, or vice versa. Similarly, new production technologies do not progressively favor certain regions over others.

Finally, I want to stress again that geopolitics does not determine foreign policy. Geopolitics simply limits the spectrum of strategic options available to a state. In fact, we cannot predict the course of a state's foreign policy by simply looking at the geopolitical variables of trade routes and centers of resources. The full range of human motivations lies behind foreign policy, and they must be taken into consideration when attempting to determine the direction of a state's geostrategy. This means that often states pursue goals that are not consistent with or informed by the geopolitical reality but are the result of ideological principles, interest-groups pressures, or other variables. In fact, states often ignore geopolitics and do not pursue control of resources and lines of communication, focusing for a variety of reasons on other, less strategic regions. Geopolitics, therefore, does not offer the motivation or the capacity to pursue a geostrategy that can reflect it.

But geopolitics constrains the spectrum of possible geostrategies. Every state, no matter how ideologically motivated, acts within a setting determined by routes and the location of resources. Control over these objectives bestows power, and even if a state is motivated by ideological concerns, it cannot ignore this cold fact. A foreign policy that does not reflect the underlying geopolitics cannot increase or maintain the power of a state.

Geostrategy

If geopolitics is the setting in which states act, what is geostrategy? And more importantly, what is the relation between geostrategy and geopolitics? Geostrategy describes the geographic focus of a state's foreign policy, or where a state directs its power. It is a descriptive and not a normative concept because it does not propose where a state ought to direct its attention and project power.

The main variable influencing geostrategy is state borders. States seek above all else to protect their territory from invasions and attacks, and state borders are a good measure of territorial security. When state borders are threatened or unstable, the state must concentrate its efforts on the preservation of its territorial security and is unable to pursue an effective foreign policy far from its territory. Diplomatic, economic, and military resources must be diverted to the protection of borders, limiting the state's ability to project power to strategically important but distant places.

Territorial security is not the same as security. In fact, because territorial integrity is only one aspect of a state's security, it can be argued that the defense of the security of a state begins far from its borders. For instance, the protection of distant resource-rich regions, such as Europe and the Middle East, or, more generally, the maintenance of a balance of power in Eurasia, has guaranteed, and continues to guarantee, the economic and political security of the United States. My argument does not contradict this idea but simply stresses the primary security importance of national borders; the protection of trade routes and resource-rich regions is an important source of power and security, but it is useless if it is accomplished at the expense of territorial security. In fact, geographically distant military commitments can deprive a state of the strength necessary to defend its home territory.[60] Consequently the defense of state borders comes before the pursuit of a geopolitically sound strategy that defends and controls trade routes and centers of resources.[61]

The stability and security of state borders are influenced by both geography—differences between land and sea borders and between the characteristics of land borders—and politics—the underlying balance of power.

The most basic differentiation is between land and sea borders. Historically, land borders have been more dangerous than maritime ones. A contiguous land power is more threatening than one separated by sea because a land invasion is operationally easier and hence more likely than an amphibious or airborne one. As a result, "border pressure" is greater on land than on sea. As Harold Sprout observed, "Science and technology have made huge strides toward the conquest of time and space. But it is still axiomatic that sea frontiers can be, and are, defended more securely, with less outlay and effort, than land frontiers. A country thus removed from other centers of military power and ambition enjoys a measure of security and a freedom of action and choice denied to less favored countries with powerful and dangerous neighbors and vulnerable land frontiers."[62]

Land borders can be differentiated by their geographic characteristics. Land borders marked by impervious areas such as high mountain ranges or heavily forested areas are difficult to cross and consequently more stable than a plain that is neither forested nor crossed by deep rivers, offering no natural barriers to invasions.

The validity of the differentiation between land and sea borders is difficult to test and is based mostly on an intuitive empirical analysis. But throughout history the success of sea-launched invasions has been mixed at best. For instance, in World War II the Normandy invasion by the Allies was a success, whereas the Nazis had previously failed to even begin a comparable invasion of Great Britain, and their airborne assault on Crete had been so costly that it was never repeated.[63]

The ambiguous historical record leads some theorists to abandon completely the distinction between sea and land borders and resort to distance as a variable influencing threat or the projection of power. This simplification has some value.

The difference between land and sea borders can be explained in part by using only distance as a variable. A land border corresponds to geographic proximity (and a greater likelihood of instability), while a sea border corresponds to geographic distance (and a greater likelihood of stability).[64] But such simplification has a limited explanatory power because, as I mentioned above, land borders do not fit neatly into one category (proximity and the resulting instability or a higher level of threat). Despite representing a smaller distance than a body of water, a mountain range might provide more security than a sea. Similarly, at the outset of World War II the English Channel provided greater security to England than the Low Countries did to France even though the distance over the sea was shorter. It is simply difficult to cross a body of water. Hence, the distinction between sea and land borders continues to be an important variable explaining the likelihood of conflict and the direction and outcome of foreign policy.

Geographically similar land borders are not all equally dangerous or stable. In fact, land borders that are easy to cross might be very stable because there is no political reason for them to be a zone of tension. Similarly, apparently impassable borders have served as conduits for invasions. As Spykman observed, "Nature alone has almost nowhere created impassable barriers" that would explain a long period of border stability.[65]

We must consider therefore a complicating factor: the politics between the states separated by the borders. The stability of the borders is a factor of the underlying political relationship between the states in question rather than vice versa.[66] The more conflictual the relationship, the more unstable the border. For instance, the U.S. home territory has been and is safe not only because it has sea borders but also because it has politically stable relationships with its neighbors and thus secure land borders.[67] Had its land borders become a source of threat, the United States would have been unable to devote resources, for instance, to the defense of Europe since World War I.[68]

From this categorization it is evident that there is a spectrum of danger to territorial security of which the borders are a manifestation. Such a spectrum, or the situation on the borders, influences the geostrategy of states by prioritizing its objectives. The first priority is the protection of territorial security. If the borders are threatened, the state will have to concentrate its efforts on the preservation of territorial security and limit the pursuit of control over resources and trade routes.

However, it is necessary to nuance this argument. Often, in fact, the protection of territorial security coincides with and does not hinder the pursuit of access to resources and routes. Unstable land borders are not a serious strategic handicap if they lie on the path to resources and routes. That is, if resources and routes are beyond those borders, their instability conflates the two goals of the state: the protection of its borders and the extension of control over routes and resources. By extending state influence to those resources and routes, the state can at the same time push the dangerous border farther from its territorial core.

The logical consequence of this argument is that a state that has long land borders is not automatically prevented from controlling important routes and resources. Contrary to what Mahan argued, not all trade routes are maritime, and consequently a sea power does not have an automatic advantage over a continental one.[69] There have been and continue to be strategic routes on land. For instance, the caravan routes through Central Asia, not the sea lanes of the Pacific and Indian oceans, were the main channels of trade between Europe and Asia through the Middle Ages until the seventeenth or eighteenth century. Today, oil pipelines are strategic routes that cross continents and are as important as the sea lanes used by tankers.

Furthermore, Mahan's argument is that the easiest way to reach resource-rich regions is by sea. Again, this depends on the location of the state in relation to the geographic distribution of resources. For instance, for the Ottoman Empire the main center of resources, Europe, was across its land borders along the Danube, not across the Mediterranean Sea. Thus, the Ottoman Empire expanded along two fronts that also coincided with its most problematic land borders: toward the main center of economic resources (Europe) and toward trade routes (along the Danube and in Persia). Similarly, more recently the strategic Persian Gulf region was within continental, not maritime, reach of the Soviet Union. Thus, some states, because of their location relative to resources and routes, do not need to become maritime powers in order to conduct a geostrategy reflective of geopolitics.

Finally, it is important to stress again that stable borders (or the coincidence of land borders and the locale of resources and routes) do not necessarily mean that the state will increase its influence over routes and resources. In other words, the stability of borders (i.e., territorial security) is a necessary but insufficient condition for a foreign policy that pursues the control of routes and resources. It creates the condition, but not the motivation, for expansion. For instance, the fact that the United States has had stable land borders since the early twentieth century has meant that it could, but not necessarily that it would, project its power to protect Europe and control the main sealanes of the world. Theoretically, therefore, a state might be satisfied with its territorial security and choose not to extend its influence and control over strategic regions outside its borders.[70] There are many reasons why a state does or does not expand, and they are not always based on a sound understanding of geopolitics.

CHAPTER THREE

The Geopolitical Change of the Sixteenth Century

At the turn of the sixteenth century the geopolitics of Eurasia turned literally inside out: Europe's Atlantic coast and Asia's Pacific shore became strategically pivotal, while the Mediterranean Sea and Central Asia became less important. This change affected the history of the Venetian, Ottoman, and Ming empires by forcing them to adapt their strategies to the new geopolitical reality. The objectives they pursued—trade routes, centers of resources—were suddenly less valuable because of competing routes and markets. In the end all three declined, in part because of misguided geostrategies, in part because of their growing irrelevance in Eurasia. The disconnect between their geostrategies and the underlying geopolitics led to their demise as great powers.

What Changed?

The voyages of Columbus and Vasco da Gama at the turn of the sixteenth century and the imperial expansion that followed altered the geopolitical situation faced by Eurasian powers in a dramatic and lasting way. As Adam Smith argued in 1776, perhaps somewhat excessively, "The discovery of America and that of a passage to the East Indies by the Cape of Good Hope are the two greatest and most important events recorded in the history of mankind."[1] These two events changed the pattern of trade routes and of power in the world. The discovery of the Americas bestowed an unexpected source of wealth on Atlantic Europe, while the circumnavigation of Africa connected Asia directly with western Europe. As a result, the continental (Central Asia, the Middle East, South Eastern Europe) and maritime (the Persian Gulf, the Black and Mediterranean seas) cores of Eurasia lost strategic relevance.

The first great change was that the main trade routes of the world moved from within Eurasia to the oceans surrounding it. The 1497–98 voyage of Vasco da Gama around Africa resulted in the discovery of a new route linking Atlantic Europe (Portugal, Spain, and later the Netherlands and Great Britain) directly with Asia, while Columbus's discovery of America in 1492 established new routes from Spain (and western Europe) to the new continent. In the succeeding decades these routes became the key lifelines of a global commercial network. The historian J. H. Parry observed that "two major systems of European oceanic trade grew up in the first half of the sixteenth century: the one between Portugal and India, specifically between Lisbon and Goa; the other between Spain and America, specifically between Seville and various harbours in the Caribbean and the Gulf of Mexico."[2]

The strategic importance of Columbus's discovery and of the resulting Spain-America route became tangible only when the newly discovered continent turned out to be a new source of wealth and power. Columbus's discovery altered the configuration of power more than it did that of trade routes, and although its impact was not as immediate as that of Vasco da Gama's voyage, it was perhaps more lasting because it introduced a new, initially unchallenged source of wealth for the Spanish Empire. Columbus discovered a new center of resources, and Vasco da Gama found a new route to a well-known center.

Trade between Asia and Europe dates back to Roman times, but it was only in the late Middle Ages that it became a key source of wealth and power. Because of the more advanced economic development of Asia, and China in particular, such trade was mostly unidirectional: Asian goods, mainly expensive but lightweight products such as silk and spices, were carried to the European markets, while relatively few European goods were in demand in Asia. The Asian goods reached Europe through a network of routes, crisscrossing Central Asia via the so-called Silk Road and following the coast of India to the Persian Gulf and the Red Sea. The land and sea routes converged in the eastern Mediterranean, from where goods were carried to the various European markets.

The link between Asia and Europe began to change sometime in the fourteenth century, privileging the maritime over the land route. There were two main reasons for this change: the growing instability of Central Asia and Vasco da Gama's expedition. The land routes through Central Asia had always been subject to the political vagaries of the region and often had been interrupted or redirected by wars and the collapse of political authority. In the early 1500s the political situation along the caravan routes deteriorated: the Timurid Empire (modern Kazakhstan) collapsed, while a unified Persia under the Safavid leadership of Ismail caused tensions with the Ottoman Empire. The resulting instability hurt commerce along the caravan routes in Asia. In fact, some historians argue that political instability in Central Asia, rather than the discovery of new maritime routes, was the main reason for the dwindling of continental trade. As Morris Rossabi writes, "Protection costs were too expensive, and plundering of cargo was a real concern. The economies to be gained from ship transport dampened still further the merchants' plans for overland trade, but a major motive for not dispatching caravans stemmed from the military and political conditions to be faced along the Asian landmass."[3]

The argument that the political chaos of Central Asia severed the commercial link between Asia and Europe is strengthened by the technical difficulties of sea commerce. Maritime shipping in fact did not offer clear advantages to Asian trade and by itself would not have caused the collapse of Central Asian routes. Silk and spices, in high demand in Europe, were light, easy to pack, and not perishable, making them appropriate for the long, slow caravans that crossed Central Asia. At the same time, in the fourteenth through sixteenth centuries maritime commerce was infrequent, seasonal, and subject to shipwrecks. All else being equal, there was no reason to choose the maritime route, whether through the known sea lanes in the Red Sea or the Persian Gulf or, after 1498, the circumnavigation of Africa, over Central Asia. It is therefore doubtful that Vasco da Gama's voyage and the subsequent Portuguese expansion in Asia were the main reasons for the shift in trade routes between Europe and Asia.

Historians have debated ad nauseam the relative importance of Vasco da Gama's expedition and of the political turbulence in Central Asia, and there is little agreement on which one has been more influential in reshaping the commercial network of routes in Eurasia. To a certain degree this debate misses the point because it seeks a monocausal explanation of the geopolitical change of the sixteenth century. Those that argue that Central Asian land routes declined on their own because of the growing instability of the region often ignore the fact that a parallel maritime commercial system had been developing in Asia since the twelfth century. At that time, in part because of the Mongol invasions, the center of economic production in China had shifted toward the coastal areas. An intricate network of maritime trade developed linking East Asia with the Indian Ocean through the Malacca Strait and linking India with the Red Sea and the Persian Gulf. According to G. V. Scammell,

> The Portuguese, on their arrival in Asia, had encountered an ancient and complex commercial network reaching by land and sea from Europe itself to China. It was far larger, and probably handled traffic of far greater value than anything known in the

West. Because of the constraints of distance and seasonal wind changes it was conducted through entrepots such as Aden, Hormuz and Melaka. It tapped alike luxurious products of China, the gold and ivory of East Africa, the cotton textiles of India and the spices of Indonesia. What little that was needed from Europe could be received via the Middle East, whilst Arabia supplied horses and Iran silk and precious metals.[4]

This maritime route competed with the caravan routes even before the early-sixteenth-century political instability in Central Asia.

Moreover, the competition between Central Asian land routes and East Asian coastal sea lanes ended in the eastern Mediterranean, where the two converged. The caravans reached it through Central Asia and the Black Sea region, the sea shipping through the Persian Gulf (and Syria and Palestine) and the Red Sea (and Egypt). Ultimately, no matter how the goods reached it, the Mediterranean was the funnel to Europe for almost all the Asian trade. The importance of the Mediterranean was not affected by the fluctuations in trade between Central Asia and the maritime route. It did not matter, therefore, whether the Central Asian caravan routes were being eclipsed by Asian coastal shipping. These two routes competed with each other, and not with the Mediterranean.

The importance of Vasco da Gama's voyage was that it directly affected the Mediterranean. His discovery of the Cape of Good Hope route and the subsequent Portuguese expansion in the Indian Ocean and East Asia opened a new terminus to the Asian trade: Atlantic Europe. Specifically, the Portuguese empire, established in the first decades of the sixteenth century, linked western Europe to Asia through a string of bases (from East Africa to Macao via Goa and Malacca) that not only created a route competing with the Mediterranean but also directly reached the source of that trade. It diverted the Asian maritime trade away from the Persian Gulf, the Red Sea, and the Mediterranean. The Mediterranean was no longer the only route to Europe from Asia.[5]

Vasco da Gama's voyage alone was not sufficient to change the trade routes. Similarly, Columbus did not single-handedly give a source of wealth to Spain. The years 1492 and 1497 were only the beginning of a long trend that resulted in what is called the "Vasco da Gama age" (in Asia), the "Atlantic hegemony," or even more broadly, the "Commercial Revolution" or the "Age of Discoveries."[6] The Portuguese expansion in the Indian Ocean, followed by the Spanish and Dutch onslaught, took decades to establish a western European foothold in Asia. And it was probably only in the late 1500s and early 1600s, a century after Vasco da Gama, that the ocean route linking Asia with Atlantic Europe replaced the Mediterranean as the principal commercial connection between these two centers of wealth.

The linking of Atlantic Europe with Asia and America marks the birth of long-distance maritime (or, more precisely, oceanic) trade.[7] Sea routes were more reliable than land ones, especially after sixteenth-century improvements in seamanship and shipbuilding allowed for regular, relatively safe trips. Furthermore, sea lanes connected regions that previously had been separated. For instance, as early as the late sixteenth century, under the auspices of the Spanish Empire, a vibrant trade developed linking East Asia with Mexico.[8] This was the beginning of a global economy.

The discovery, development, and management of new oceanic routes allowed the creation of a commercial network that was larger than the one based on land routes (and internal maritime routes, such as the Mediterranean).[9] It is true that like land routes, sea lanes had to be maintained by a power to ensure free and safe passage to trade, but it was cheaper to maintain a maritime commercial network of ports than to impose imperial control over vast land areas. The Portuguese, for instance, preserved their sixteenth-century Asian empire from East Africa to Macao with only ten thousand troops. As Debin Ma comments,

> The nature of the open sea meant that the survival of long-distance trade no longer depended solely on the shifting political cycles of giant land-based empires. So long as traders had enough power to fend off seaborne piracy, they could bypass intermediaries and trade directly with destination port cities though all-sea routes. . . . The cost of keeping sea routes open and safe for lucrative long-distance trade—the suppression of seaborne piracy and the securing of strongholds at strategic trading ports—was much lower than that for controlling overland routes, which normally required military conquest and administration of alien territories.[10]

Finally, da Gama's voyage changed not only the configuration of trade routes, the amount of trade, and the medium of commerce but also the actors involved. The direct link between western Europe and Asia (and America) transferred the seat of power in Europe westward and altered the constellation of power in Asia. It bestowed strategic importance to those who had access to oceanic routes, decreasing the commercial and political value of the land routes in Central Asia. To use Nicholas Spykman's terms, Eurasia's Rimland became more important than the Heartland. As William McNeill wrote, "European ships had in effect turned Eurasia inside out. The sea frontier had superseded the steppe frontier as the critical meeting point with strangers, and the autonomy of Asian states and peoples began to crumble—exposed, as they were, to European armies and navies equipped with ever more formidable weapons and managed by increasingly effective national governments."[11]

This change affected the strategic calculations of the established powers of the fourteenth and fifteenth centuries. In the Mediterranean Sea, Genoa, Venice, and the Ottomans suddenly had to face a radically different situation: they became commercially expendable and had to compete with new actors for control over European trade, which until then they had considered to be under their quasi-monopolistic control.[12]

In Asia the era of Vasco da Gama brought perhaps even greater geopolitical changes than in Europe. Not only did it lead to a realignment in favor of the coastal regions of South and East Asia but it introduced new powers into Asia. The European powers, led by Portugal, expanded and controlled maritime Asia, forcing the local actors to accept their hegemony. This was a geopolitical upheaval of enormous consequences that, in different form and with different actors, continues to characterize Eurasia. As Scammell observes, the "arrival of da Gama was to mark the end—although this was hardly apparent at the time—of Europe's subjection to those incursions from the East which it had endured since Antiquity. It likewise heralded the beginnings of western hegemony in Asia." Moreover, Asia became a theater of European conflicts. "The East was sucked into European rivalries, became central to European grand strategies, and was often the victim of forces originally deployed by one western power against another."[13] It was a new geopolitical situation that, as we shall see, forced Asian (e.g., Ming China) and European (e.g., Venice) powers to reconsider their entrenched strategies.

Thus, the geopolitical change of the late fifteenth and sixteenth centuries was momentous. The geopolitical reality of the years 1000–1500 had been characterized by the Mediterranean, the Middle East, and Central Asia serving as intermediaries between Europe and Asia, with Eurasia's seat of power at its geographic center. At the turn of the sixteenth century the connection between Europe and Asia, and with it the seat of power, moved to the oceans surrounding the Eurasian landmass.[14]

Why Was There a Change?

Geopolitical shifts are multicausal. Because of their very long time frame, such changes are affected by multiple causes that work at different moments in history and on different levels. Geopolitical changes are, as John Lewis Gaddis called them, tectonic shifts resulting from several forces, none of which alone could, or had the intention to, change the geopolitics.[15] First and foremost this means that there was no grand strategic plan on the part of a king or a state to alter the geopolitical situation. No state or individual can plan and implement a change in geopolitics. These are processes that are too complex and large to be managed by one state. And they are often the unexpected result of small decisions. For example, Portugal began the exploration of West Africa and later of a new route to Asia more out of curiosity than out of a desire to change the geopolitical situation. Even if, as some historians have argued, Portugal wanted to find an alternate route to Asia in order to avoid Ottoman-controlled territories, it did not foresee the geopolitical consequences of the discovery of the Cape of Good Hope. The Spanish discovery of America was even more uncontrollable and unforeseeable. Columbus and the Spanish royals supporting him wanted to find another route to Asia and instead found a new continent, which brought Spain greater wealth and strategic weight than a direct link to Asia would have brought. Therefore, although they are easier to individuate as causes of geopolitical changes, the decisions and actions of individual statesmen and states have an effect on geopolitics that is rarely calculated and willed.

Moreover, any single action is not enough to alter geopolitics. Columbus's discovery or Vasco da Gama's voyage had to be followed by decades of further exploration, conquest, and management and defense of the newly acquired territories. Their individual actions started a trend that only after a considerable period of time, several decades at least, resulted in a geopolitical change. The discoveries of the late 1400s resulted in a geopolitical change only in the late 1500s. During the course of that century the geopolitical change—the switch from the Mediterranean to the Atlantic, from Central Asia to the Indian and Pacific oceans—was only one among many possible outcomes of the original discoveries. At any given point Portugal and Spain could have decided not to pursue their imperial expansion in the Atlantic and in Asia, and the geopolitical change initiated by Vasco da Gama and Columbus would not have occurred.[16]

The difficulty of explaining this geopolitical change has led many historians and social scientists to seek more abstract causes based on the premise that large processes need large forces. The question also becomes broader. It is no longer Why Vasco da Gama, not Cheng Ho? but Why Europe, not Asia? Why Portugal, not China?

The answers are various and controversial.[17] Here I will limit myself to two broad categories of explanations: the geographic and the technological. The geographic explanation is straightforward: Europe, in particular western Europe, had geological features that made it easier to develop a vibrant economy based on commerce, to exchange ideas and technological innovations, and to have a more productive agriculture and industry. I have examined the gist of such explanations in previous chapters. Here I want to point out only the most interesting explanation of this category, which argues that Europe had an advantage over other continents, especially Asia, because it was subject to fewer environmental disasters. Disasters are defined as "abrupt, major, negative shocks which reduce the aggregate assets or income of a given population." The growth differential between western Europe and Asia and the consequent geopolitical change can be explained through a simple analysis of such negative shocks. Europe grew faster and became a greater power than Asian because "Europe is and was a safer piece of real estate than Asia."[18]

Geographic explanations have several limitations. As previously noted, they border on determinism. Moreover, geography does not explain variation in history. Specifically, a purely geographic explanation does not explain why Atlantic Europe was weaker than the Mediterranean powers and, arguably, than China until the sixteenth century.[19] In other words, these explanations focus on the relative decline of Asia in the fifteenth and sixteenth centuries and the rise of western Europe and do not account for the remarkable success of China and Asia in general before the Vasco da Gama era.[20]

The second explanation of the geopolitical change in the fifteenth and sixteenth centuries has to do with technology. Atlantic Europe had an advantage over the Mediterranean and Asian powers because of its more advanced, mostly military technology.[21] And in the moment of the encounter between Europe and Asia (and between Europe and America), the Europeans had simply better technology, from artillery to tactics and military organization.[22]

European superiority was particularly evident in naval technology. This superiority allowed Portugal and the other Atlantic powers to project power farther and with smaller expenditure of manpower than, for instance, China, which had to devote its resources to controlling its continental borders. As McNeill points out, "Supremacy at sea gave a vastly enlarged scope to European warlikeness after 1500" because it allowed Europeans to reach and control from a distance other parts of the world.[23] Moreover, because of their naval superiority European powers could avoid costly continental wars that would have been necessary to control the vast territories in Asia. On land European technological superiority was offset by a marked imbalance in manpower in favor of Asian states.[24] Europeans, the Portuguese in particular, placed their bases where there was less power to threaten them or where they could leverage local divisions to maintain their control over the city or port. They rarely expanded past these bases, for instance, inside China or in the Middle East, where there were powerful empires that would have required an

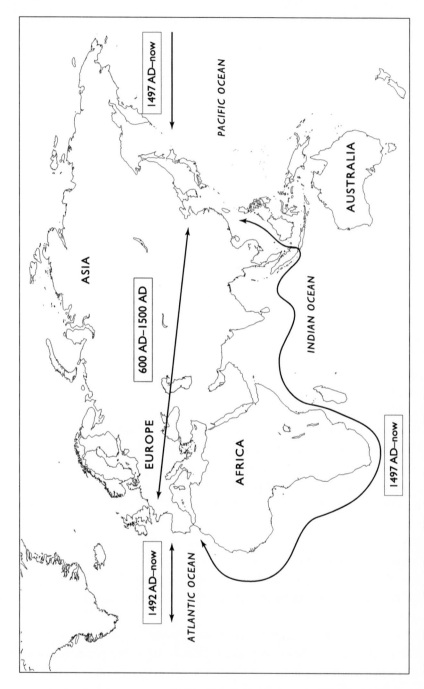

Map 1. *The geopolitical change of the sixteenth century: old and new routes*

enormous expenditure of military strength and manpower to battle. As a result, "colonial wars were . . . 'small wars.' "[25] It was only in the eighteenth century that Europeans conquered large parts of continental Asia.[26]

The main challenge to European powers in Asia was other European states in search of wealth. When European empires began to compete for the control of Asian sea lanes, military expenditures rose. It was more expensive for the Dutch to dislodge the Portuguese from Asia in the seventeenth century than it had been for the Portuguese to conquer key strongholds in the region a century earlier.[27]

European superiority in naval technology can be summed up by three developments: the introduction of the compass, improvements to maps, and improvements to ships. The introduction of the compass, combined with increasingly more precise ways of fixing the latitude and longitude of ships, allowed pilots to navigate in unknown waters and far from coastal landmarks. Vasco da Gama's voyage around Africa, for instance, was remarkable because it followed the prevailing wind patterns far away from the African coast in the Atlantic and close to the Brazilian shores. Chinese and other Asian ships (as well as, to a smaller degree, Ottoman and Venetian ships) were unable to perform such feats of navigation.[28]

In the fifteenth and sixteenth centuries maps were becoming more detailed and reflective of reality. The first maps, printed in Europe in the 1470s, had been a sign of authority and a source of power.[29] As Geoffrey Parker has observed, "Maps became for the first time a standard instrument of government—a vital tool both for mobilizing the state's resources at home and for projecting its power abroad." And a "government that lacked the cartographic tools required to organize its resources or to project its power, and instead resorted to outdated general atlases for strategic planning, was no longer a convincing imperial power."[30]

Finally, perhaps the most visible example of European naval superiority was in shipbuilding, specifically in sturdier and oceanworthy vessels. Beginning in the fifteenth century the Portuguese and in general all Atlantic navies were much more powerful than those of the Mediterranean and Asian powers. One possible explanation is geographic: the geographic environment in which ships had to function forced shipbuilders to adapt quickly. The Mediterranean Sea and, to a degree, the Indian Ocean required lighter ships capable of traveling through shallow and often windless waters. The preferred and most convenient way of sailing was by following the coast and hopping from port to port. For instance, in the Mediterranean the galley was the main type of ship: its large crew served as oarsmen to maneuver the ship in coastal waters and as soldiers in naval skirmishes.[31] While the galley's construction made it easy to pilot in coastal waters and windless seas, it also required frequent stops to replenish food and water supplies.

The design of the Atlantic ship was based on the need to travel greater distances without the possibility of anchoring in a safe harbor. This required not only a sturdier vessel but also a smaller crew, which in turn meant that the ship could be out of port for longer periods of time, not needing large quantities of food and water. The resulting ship was heavier (and thus capable of carrying greater artillery firepower) and sail rigged (and thus faster and able to travel greater distances). The superiority of Atlantic shipbuilding became evident once guns were put aboard ships: the Mediterranean galleys and the Asian ships were too light to be able to withstand the weight and recoil of the guns. The 1588 defeat of the Spanish Armada by the English navy was the defeat of the Mediterranean (and generally speaking of the non-Atlantic) ship and naval tactics.

Thus, in the sixteenth century the Eurasian Rimland—Atlantic Europe and East Asia—became strategically more important than the Heartland—Central Asia and the Mediterranean Sea. A combination of factors, from geography to technology and the actions of a few bold explorers, contributed to this geopolitical change. This was the geopolitical situation to which Venice, the Ottoman Empire, and Ming China had to respond.

CHAPTER FOUR

The Geostrategy of Venice (1000–1600)

On Ascension Day, 9 May 1000, Doge Pietro Orseolo II led a fleet of ships from the Venetian lagoons toward the Dalmatian coast. Constantly harassed by pirates based around the mouth of the Narenta River, the Slavic populations of Dalmatia happily accepted the fleet and the authority of Venice. The Venetian galleys annihilated the pirates' fleet and strongholds. In the ensuing weeks the doge established control over Istria and a string of cities along the eastern coast of the Adriatic Sea. Venice became the protector of Dalmatia, and the doge assumed the title duke of Dalmatia. The seeds of the Venetian empire had been sown.

In the succeeding decades Venice extended its influence from the safe lagoons at the western edge of the Byzantine Empire to the distant regions of the eastern Mediterranean and the Black Sea. The Republic of St. Mark "married" the sea, dominating it from the lagoons to the Aegean.[1] By the end of the thirteenth century the Venetian empire was perhaps the greatest Mediterranean power since Rome. Venice exercised a monopoly over trade between Europe and Asia, defeating potential competitors and befriending commercial partners. At the end of the fifteenth century, however, Venice had started to decline and soon became irrelevant in the international arena. While its independence officially ended only in 1797, at the hand of Napoleon, Venice had lost its luster more than two centuries earlier.

The rapid rise and the protracted decline of Venice raise important questions. Why did it develop as a great power in the Mediterranean? Why at that moment in history? And why did it become strategically irrelevant? These questions are important because the responses illustrate the relationship between geopolitics and geostrategy.

Venice rose to the rank of a great power because of its location and the foreign policy it pursued. Venice was situated at the end of a key route from the East, in the northern Adriatic. It was the access point to the European markets for Eastern wares. Location alone, however, did not suffice to make Venice a great power. In fact, Venice directed its foreign policy and developed its system of bases and allies eastward, following clear geographic objectives made salient by existing naval capabilities. An uninterrupted string of bases was required to make sea journeys safe and regular. Finally, Venice declined because of a misguided geostrategy and a radical change in the underlying geopolitical reality. The strength of a geostrategy depends on its mirroring the geopolitical reality. When the strategy diverges from that reality or when the reality changes without a corresponding change in the strategy, the state loses its influence in international relations. Such a divergence between geostrategy and geopolitics made Venice a wealthy but irrelevant city in northern Italy for the last two centuries of its independence.

Why Venice?

The history of the Venetian empire presents a puzzle for its students. In fact, despite the length of its history and the reach of its imperial possessions, Venice has rarely been studied in the field of international relations. There are three reasons for this omission. First, a large part of the historical literature is tainted by a strong bias in favor of Venice. Venice is too often viewed in quasi-mythical terms as an exemplary republic with a rare and early constitutional arrangement, a view made even more romantic by the dazzling artistic beauty of its islands. Venice is presented as La Serenissima, "the most serene," a romantic city that controlled half of the Mediterranean while being a refuge from the confusing and violent politics of the Italian and the European mainland. The problem with such a view is that it judges this imperial city based more on its appearances than on its political and strategic achievements.[2] This is not an insurmountable problem, but a good dose of skepticism is required when combing through the relevant literature.

The subsequent course of European history is also to blame for the lack of a continuous academic interest in Venice. The strategic successes of Venice have been eclipsed by the later global empires of Portugal, Spain, Holland, and above all Great Britain. Similarly, the vastness of the preceding Roman Empire and the coeval Byzantine and Holy Roman empires make the relatively few islands under the Venetian administration seem puny in comparison. Venice's direct control remained limited to the eastern Mediterranean and the Adriatic Sea, with commercial relations with western Europe and Asia. However, the small territorial expanse of the Venetian empire does not properly reflect its power. A strategic advantage and great-power status can be achieved by controlling a few small but pivotal places. By holding a string of strategically located ports, Venice controlled not only half of the Mediterranean Sea but also the main trade route between Asia and Europe, an achievement that puts this city on the level of the preceding or subsequent empires. Indeed, Venice was considered by its very contemporaries as one of the greatest powers, a status that was even more impressive because of Venice's independent position between western Europe and Byzantium. By the late twelfth century Venice was independent of both the pope and the emperor and was considered a power equal to other Italian and European states. The Peace of Venice, brokered by the doge between Pope Alexander III and Emperor Frederick I Barbarossa after the battle of Legnano in 1176, illustrates the early political influence of this city.[3]

Finally, it can be argued that lessons drawn from the history of Venice are no longer applicable. The geopolitical reality faced by Venice was determined to a large degree by the technologies of that age, and in the era of nuclear ships and cruise missiles the lessons of that period are irrelevant. For instance, Venetian galleys required a string of closely spaced ports where they could replenish their stock of food and water and seek refuge in case of bad weather. Modern ships, on the other hand, can stay away from ports for months at a time. Some have attempted to salvage the lessons of Venice and of that period by arguing that galleys had navigational characteristics similar to those of steam vessels.[4] But even if they were technologically similar, the similarity was not strong enough to argue in favor of the relevance of Venetian history for modern politics. The historical period was unique in many ways, especially in terms of technology, and it is dangerous and useless to draw tactical lessons from this case. However, the case of Venice, like the succeeding cases, offers important lessons on the strategic level. For instance, in light of Venice's technological capability and the geopolitical environment, its formulation of a grand strategy, as well as its failure to formulate such a strategy in the second half of the fifteenth century, offers an important lesson that transcends the historical specificity of Venice.

The last preemptive defense of this case and of this approach has to do with alternative explanations. Why are geography and geopolitics, and not other explanations, important? I do not claim that geography explains completely the grand strategy of Venice or the reasons for its decline. The commercial and adventurous spirit of the small Venetian population, their capacity to increase their wealth rapidly, their renowned diplomatic genius, and the stability of their domestic institutions all help to explain the success of this empire. Similarly, the weakening Byzantium, coupled with the confusing political situation in Italy, provided the setting for Venice to develop and expand. However, in order to explain the foreign-policy direction undertaken by Venice, along with its political success and its silent decline, it is not enough to look at the internal development of Venice or Byzantium or simply at the constellation of powers. Venice expanded in a clear direction, toward the eastern Mediterranean, avoiding nearer and wealthier territories in Italy or Central Europe. I argue that this direction reflected correctly the underlying geopolitics. The geostrategy of Venice, with its network of bases and allies located along key trade routes, allowed it to endure the attacks of other powers (Byzantium, Genoa, Ottomans) that often deprived it of some portions of its empire without succeeding in conquering it.

There are also valid alternative explanations for Venice's decline, namely, the rise of large and efficient empires in western Europe toward the end of the fifteenth century (France and Spain). These states presented formidable challenges to Venice, which had to face their intrusions in Italy and the Mediterranean. Arguably, the Republic of St. Mark was simply too small to compete with the well-organized western powers, which managed to harness resources through taxation and turn them into impressive military capabilities. Venice's expansion on the Italian mainland could be interpreted as an attempt to respond to its lack of a taxable territory, a weakness that was evident when the city had to field armies capable of standing up to those of France, the Habsburgs, or the Ottomans.[5] I do not deny the validity of this explanation. The internal structure of Venice, from its fiscal system to its political constitution, was very different from that of the other European (but not Italian) states and may have affected Venice's ability to face them. Nonetheless, the wealth Venice derived from its maritime empire was also enormous and, had it not been disregarded from the fourteenth century on in favor of Venetian mainland possessions, could have continued to maintain the position of superior power enjoyed by the republic. I argue here that Venice faded not simply because it was stuck with a less efficient state model but also, and perhaps most importantly, because the source of its wealth and power, the maritime commercial empire it built in the Mediterranean, was neglected and then lost its vigor owing to a change in routes. Any explanation of Venice's decline that ignores this source of its strength is bound to be incomplete.

Two questions inform this chapter. First, where was Venice? The position of Venice on the Eurasian map was crucial to its grand strategy. Second, how did Venice respond to this map? I examine why and how Venice expanded in the eastern Mediterranean. I divide Venice's history into three periods: the rise to power (1000–1204), the management of its empire (1204–1400), and the decline of its influence (1400–1530). Because by the mid-sixteenth century Venice was no longer a noteworthy power but merely a splendid postimperial city, I do not examine the last two centuries of its independence.

The Geopolitical Reality

From the eleventh to the fifteenth century Venice was located on the main route linking Europe with Asia, and it used this location to its benefit.

The Location of Resources
NATURAL RESOURCES: SALT, TIMBER

Economic and natural resources attract the attention of states, which compete to secure their supply.[6] Venice had two distinct advantages in this field. First, it had easy access to two vital natural resources, salt and timber, and second, it was located on the geopolitical frontier between western Europe and the Byzantine Empire (or, later on, between western Europe and the Ottoman Empire), with access to these centers of resources.

Salt production was a key source of government revenue. Because salt was found in abundance near Venice, the Republic of St. Mark quickly became one of the principal suppliers of salt to markets in northern Italy. Soon the Italian markets consumed more salt than Venice could produce in its lagoons.[7] By the thirteenth century, in part because of the desalinization of the lagoon caused by rivers, Venice started to import salt from Cyprus and even the Balearic Islands. Moreover, it maintained the monopoly of salt production by compelling potential competitors such as Ravenna and Cervia to sell their salt only to Venice. At the same time it signed treaties with buyers forcing them to accept Venice as their only supplier.[8]

Access to timber was indispensable for building and maintaining merchant and military fleets, but several centuries of heavy exploitation had reduced the Mediterranean region to an almost treeless landscape.[9] Venice had the advantage of being located near two major sources of timber. One was around the lagoons and, a little farther away, on the slopes of the Alps; the other was the Dalmatian coast. As the forests around the lagoon became depleted, Venice had to extend its control increasingly farther to maintain a strategic edge over the other Mediterranean naval powers.[10] The unhampered access to timber allowed Venice to be a leading shipbuilder in Europe until the fifteenth century, when, after nearby timber supplies dwindled, it had to compete with other cities, such as Ragusa, on the Dalmatian coast.[11]

THE GEOPOLITICAL SITUATION: BETWEEN EAST AND WEST

Thus, salt and timber gave Venice an initial edge over other powers. Nonetheless, resources alone did not suffice to build the foundations of the Venetian empire. A key factor, in fact, was the geographic disposition of the centers of resources, or markets, from the tenth to the fifteenth century. The Northern part of the Adriatic was situated between two distinct centers of economic resources. By serving as an intermediary, Venice benefited from this geopolitical situation more than from the direct control over natural resources, gaining not only immense wealth but also political leverage over the great powers of the time.[12]

From its political beginnings in the eighth and ninth centuries until the Fourth Crusade in 1204 Venice was the westernmost territory of Byzantium. A cursory look at a political map of the eleventh century shows that the Byzantine Empire, already in severe political decline, had only a few footholds in Italy. The 1054 schism weakened Byzantium by diminishing its connection with the Western Empire and with Europe in general. In light of the subsequent political consolidation and economic rise of Europe, this separation contributed to Byzantium's decline. Furthermore, the schism generated inimical feelings toward Byzantium in Rome. Between 1060 and 1070 the pope supported the Norman leader Robert Guiscard's engagement in a sweeping campaign conquering southern Italy and thus ending Byzantine influence on the peninsula.[13] In 1071 Bari fell, leaving Venice as the only Italian ally of Byzantium and, most importantly, as its only access point to European markets. The Muslim attacks in Asia Minor, accompanied by troubles in the Danube region and the rise of pirate activity on the Dalmatian coast, compelled Byzantium to rely increasingly on Venice for both military protection and commercial support. The Venetian expedition to Dalmatia in 1000 was the first significant sign of Byzantine decline and of Venice's shift from a Byzantine protectorate to an ally and then to a challenger.

The decline of Byzantium, however, did not end the Eastern center of resources. In fact, in the thirteenth and fourteenth centuries the eastern Mediterranean was a thriving region characterized by a mosaic of political entities, such as the "Latin empire" in 1204 and the crusaders' states in Palestine. The Crusades and the resulting political entities in Palestine were a thriving business, enriching the maritime powers of the time. A continuous flow of pilgrims and soldiers from western Europe to the liberated cities in Palestine was accompanied by growing commercial opportunities in the eastern Mediterranean.

The rise of the Ottomans, who expanded in Asia Minor until completely surrounding Constantinople in the early fifteenth century, slowed only momentarily the growth of this region. The slow but irresistible ascent of the Ottoman power, with its final conquest of Constantinople in 1453, unified the eastern Mediterranean again. The political change was momentous: the Eastern Roman Empire ceased to exist, while the Ottomans controlled Asia Minor and Palestine and expanded in the Aegean and Black seas. Orthodox Christianity, indeed Christianity as a whole, seemed to be mortally threatened. However, from a geopolitical perspective the rise of the Ottoman Empire did not result in the collapse of the Eastern center of resources. The fact that despite their naval strength and Christian character Venice and Genoa did not help Constantinople against the Ottomans illustrates well the continuity of the geopolitical situation. The Ottomans were open to Western merchants, and even before the fall of Constantinople Venice had a well-developed trade through Ottoman-controlled ports. Because Venice never strictly obeyed pontifical decrees a change in the governance of the eastern Mediterranean did not affect Venetian fortunes.[14]

The reason for the geopolitical continuity in the eastern Mediterranean was that the Byzantine, Latin, and Ottoman empires were not only a source of commercial goods in themselves but also the most important intermediaries for products coming from Asia. As long as the authorities in control of the eastern Mediterranean were supplied with, and allowed the passage of, Asian wares they would attract the attention of European powers.

On the western side of the geopolitical map, the situation from the eleventh to the sixteenth century was characterized by political confusion but economic growth. By the end of the tenth century the Western Roman Empire was trying to extend its power over the papacy, thereby strengthening the political foundation of the empire. Nonetheless, the imperial attempt to unify Italy and Europe failed. Five years after the 1054 schism between Rome and Constantinople Pope Nicholas II decreed that all subsequent popes were to be elected by a council of cardinals, depriving the Western emperor of the power of investiture. Later, Pope Gregory VII excommunicated the Emperor Henry IV, who in 1077 had to beg for forgiveness in Canossa. But even the strengthening of the papacy was temporary (as made clear by Henry IV's return to Italy a few years later and by Gregory VII's flight from Rome in 1089), and the conflict between the pope and the emperor, and between their Guelf and Ghibelline supporters, continued throughout the fourteenth century.

Paradoxically, the political confusion of these centuries had beneficial effects on European growth. Between the eleventh and fourteenth centuries the population of Europe doubled, perhaps even tripled, because of climatic improvement, agricultural revolution, and infrequent plagues, among other factors. Production increased, fueling exports of manufactured goods, such as cloth.[15] The economic development in this period was so great that the late Middle Ages are often referred to as the "Twelfth Century Renaissance."[16]

The struggle between the secular and religious powers over the political independence of the pope weakened the imperial hold on Italian and, to a smaller degree, northern European cities. Starting in the last decades of the eleventh century several cities obtained different forms of self-government from the empire: Pisa in 1075, St. Omer in 1127, followed by Bruges, Ghent, and Lubeck in 1143 and Hamburg in 1189. In 1082 Venice had obtained similar benefits from the Eastern Empire. The charters granting independence to these cities led to an economic renewal of Europe.[17] Annual fairs in six cities centrally located in the French region of Champagne illustrate well the rapid economic and commercial growth of Europe. Merchants would converge from all over Europe to exchange wares and, in an incipient form of banking, pay off or obtain loans. These fairs, each lasting six weeks and not overlapping the others, allowed for commerce to continue uninterrupted almost all year.

The centrifugal forces of the Holy Roman Empire, exacerbated by the conflict with the papacy, allowed for the rise of new political actors in Europe. The authority that had been dispersed by foreign invasions and the decrepit state of the empire was being "reconcentrated" in the hands of kings and princes.[18] The kingdom of France, fueled by rapid demographic and economic growth, especially in its northern provinces, took shape in the eleventh and twelfth centuries, quickly becoming an important political and economic player on the European stage. In the fourteenth century France replaced the Holy Roman Empire as the predominant power in Europe.[19]

In Italy the decline of the Holy Roman Empire and of the papacy led to the flowering of cities such as Florence, Milan, Genoa, and Venice. An extremely fluid and complicated balance of power linked the numerous political actors, often locking them in prolonged conflict. The constant interventions of the Western Roman Empire and later of France could not but exacerbate the bitter struggles among the cities. Nevertheless, the continuous struggles on the Italian mainland also illustrate the great wealth of the various cities that could afford long conflicts without becoming exhausted. The population of the Italian peninsula, for example, increased from 5 million at the beginning of the eleventh century to almost 10 million before the plague in the fourteenth century.[20]

Over the course of the five centuries examined here the nucleus of power in Europe gradually shifted westward, toward northern and Atlantic Europe. This geopolitical shift did not become clear until the late sixteenth century, when the power of Spain, Holland, and then England dramatically altered the map of Europe. At the same time, to the east of Venice the Byzantine Empire was replaced by the Ottomans, and this region continued to be an important center for Asian goods, highly desired in Europe.

Trade Routes

The patterns of trade routes were determined by the geographic distribution of resources described above. The East-West axis connecting the eastern Mediterranean (and beyond it, Asia) with Europe was a fundamental conduit of goods and force during the Middle Ages.[21] The line of communication between Europe and Byzantium (and later on, the Ottoman Empire) was one of the most vital trade routes, fueling the renewal of European commerce. Trade with the Levant supplied Europe with luxury goods, such as wine and Asian spices and silk, which were in high demand in the increasingly wealthy cities of northwestern Europe and Italy. Cretan wine, for instance, was in high demand as far away as England.

The axis between Europe and the eastern Mediterranean (and Asia) took the concrete form of routes used for trade and, when necessary, the projection of military forces. The configuration of routes was determined by communications technologies and by the political circumstances affecting potential itineraries. Navigational skills and the range of ships, for instance, determined the positions of indispensable resupply bases and preferred courses.

Furthermore, the importance of ports in the eastern Mediterranean was determined by the flow of Asian goods; when such flows shifted because of changing political circumstances in Central Asia or Persia, the importance of ports in Palestine, Egypt, or the Black Sea fluctuated. For example, the swift expansion of Genghis Khan's hordes in Central Asia and eastern Europe made the land transport of Asian goods to Europe precarious and unreliable. Starting in 1207 the Mongols conquered southern Siberia, moving into Transcaucasia and today's European Russia and Ukraine (1230–1240s). Throughout the thirteenth century they made devastating forays as far as Cracow in Poland (1241) and Fiume in Dalmatia, finally settling on the Volga River. The result of these conquests was that the Central Asian caravan routes linking China with the Black Sea and modern Ukraine were severed. The alternate routes that crossed Persia and the Red Sea and reached the eastern Mediterranean ports became the preferred lines of communication with Asia. The benefits of the Mongol expansion for the Mediterranean, and for Venice in particular, were so substantial that some historians insinuate that the Mongol attack on Kiev in 1240 was undertaken with Venice's active encouragement.[22] Even after the Mongols settled down, stabilizing their empire, the overland routes through eastern Europe were used infrequently because of the competing cheaper sea routes through the Black Sea and the Mediterranean.

The routes from Asia to Europe had three main termini: the shores of the Black Sea (Crimea and Trebizond), in Syria and Palestine (Tripoli, Beirut, Tyre,

Acre, Haifa, Jaffa), and in Egypt (Alexandria). Two overland routes converged at the first terminus on the Black Sea: a northern path from China ending near the Azov Sea and a southern route from India through Persia ending in Trebizond.[23] The reliability of the Black Sea as a route depended on the political conditions of its shores, infested by the Mongols in the north and by Turks in the south. Also the geography of the Black Sea granted enormous strategic leverage to the power in charge of the Bosphorus and Constantinople, the choke point of this trade route. As we will see, Constantinople was a prized objective for the various powers interested in Black Sea trade.

The second route was from India through Persia and Syria, ending on the Syrian-Palestinian coast. There ports such as Tyre, Jaffa, and Acre thrived during the Crusades, especially during the Latin Kingdom of Jerusalem (1099–1187). Nevertheless, even after the fall of Jerusalem the remainder of the crusaders' possessions, mainly the fortified port of Acre, continued to play an important commercial role. In fact, Acre was such a vibrant trading port that in the second half of the thirteenth century the Muslim Mamluks were ambivalent about conquering this city, fearing that its destruction would deprive them of a vital outlet for their goods.[24] When Acre finally fell in 1291, the terminus of this route moved north to Lajazzo in Asia Minor or Cyprus.

Finally, a sea route connected India and China with the Mediterranean ports via the Red Sea. Asian and Muslim merchants brought their wares to the Red Sea, stopping at Jiddah and either continuing by land to Mecca (which in and of itself was a profitable location because of Muslim pilgrims) and ending in Syria or continuing via sea to Suez and crossing a brief stretch of land to Alexandria.[25] Despite Muslim control and the pontifical order not to trade with the "infidels," this was a popular trade route for European, especially Venetian, merchants.[26] From these ports ships "hopped" from one Aegean island to another, with Cyprus and Crete as major bases, to reach the European mainland. By hugging the Greek coast, ships entered the Adriatic or continued to the Tyrrhenian Sea.

The Adriatic Sea was a particularly important avenue of commerce for two reasons. First, the northern end had easy access to the internal markets of Italy and the transalpine routes to Europe. The rivers Po and Adige allowed merchants to travel by ship as far as Lombardy, bringing them closer to the Alpine passes through which they could continue their travel.[27] The route through the Tyrrhenian Sea was more difficult. Geographically Amalfi and Genoa, the two main competitors of Venice, had very limited access to the Italian mainland and from there to European markets. Mountain ranges separated them from the internal regions, making trade more difficult and costly.

Second, the Adriatic was under the benign surveillance of Byzantium until the eleventh century and of Venice from then on. This almost uninterrupted control kept the Adriatic free of pirates, rendering sea travel reliable and safe. The Tyrrhenian Sea, on the other hand, was infested with Saracens pirates, who also controlled Sicily, making the westward passage of goods risky and limiting trade to regional exchanges.[28] Only in the fourteenth century did Genoa manage to overcome this political limitation, transforming the Tyrrhenian into a safe trade route and competing with Venice.

Control over the Adriatic was hinged on the straits between Otranto in Italy and Corfu in Greece (and Durazzo in Albania). The power that controlled these straits closed, and thus controlled, the Adriatic. As long as that power was favorable to the northern Italian and European powers, the majority of merchants preferred to travel through the Adriatic.

Finally, within the Adriatic the preferred route was along its eastern shore. The navigational skills of the period were still very primitive, and ship pilots were wary of sailing without maintaining visual contact with the land. Ships hugged the coast, seeking harbor at night and in rough weather. Moreover, the prevailing type of ship, the galley, while useful in the often windless Mediterranean waters, required frequent stops to resupply its large crews of oarsmen with food and water. The geographic features of the Dalmatian coast were perfect for this kind of navigation. The shore is broken by inlets and protected by scattered islands, all offering safe anchorage to ships. On the western side of the Adriatic, in contrast, the coast has no "natural safe refuges, it is not backed by high mountains which would assist coastal navigation by landmarks, there are few offshore islands to provide shelter in bad weather, the holding ground is generally poor, and there are large extents of dangerous shallows."[29] The Italian coast is more linear, with few natural harbors, and captains were notoriously afraid of sailing near the shore out of fear of beaching their ships.

Venice's Borders

The city of Venice, the center of its empire, benefited from a secure geographic position. This security allowed Venice to focus its attention and resources on distant areas in the eastern Mediterranean rather than on protecting the homeland.

Rialto, the main island of the city of Venice, is separated from the mainland by a few miles of water.[30] Despite its shallowness, this body of water sufficed to protect the islands from assaults from the mainland. During the period examined here no power had the necessary amphibious capability to invade the island.

Moreover, in the lagoon the "secret and narrow" channels created by tidal currents and silt deposited by rivers were easy to block, and expert pilots were required to guide heavy ships to and from Rialto.[31] For instance, Venice survived and won the 1378–81 conflict with Genoa, the so-called War of Chioggia, by clogging these channels, thus cutting the Genoese contingent on the southern coast of the lagoon off from its naval support.[32] On the eastern side, the *lidi*, long sand bars, protected the lagoon from storms and from seaborne assaults from the Adriatic Sea. During the course of history the Venetians reinforced these sand bars by planting trees and building forts.

The security of the lagoons attracted Venice's first inhabitants. In the sixth through eighth centuries foreign invasions created havoc in the already declining Roman Empire, forcing the mainland population to seek refuge. The islands in the lagoons were a natural haven that, while close enough to the mainland to supply food, were securely distant. And they remained untouched until the loss of Venice's independence at the end of the eighteenth century.

Furthermore, the insular location allowed Venice to be detached from the confusing, often deadly medieval politics of the Italian mainland. Venice kept its involvement in mainland politics to a minimum. It had to exert some control over adjacent *terraferma* in order to meet its objectives, which included securing control over access to the internal Italian markets, preventing the rise of naval competitors in the Adriatic, and guaranteeing an uninterrupted supply of foodstuffs to the Venetian islands. For example, Venice extended its borders to the estuaries of the Po and Adige rivers because they were important access routes to the internal Italian markets. Similarly, some mainland territories were necessary to ensure that Venice had sufficient supplies of food and fresh water.[33] Moreover, Venice maintained some control over cities like Ravenna and Cervia to prevent them from competing in the shipbuilding and salt industries.

Until the first half of the fifteenth century these mainland possessions remained limited to a few strategically located places. Venice was happy to keep Italian powers, such as Milan, Florence, and the Pontifical State, at arm's length by skillfully playing the balance-of-power game, but without getting entangled in stable alliances and, above all, without fighting on the mainland.

This low "border pressure" allowed Venice to concentrate its resources on distant regions. A grand strategy that transcended the lagoons, the Adriatic Sea, and even the Mediterranean would not have been possible without this fundamental geopolitical reality. As I will show, when the border pressure increased because of Venice's expansion on the Italian *terraferma*, the republic was forced to curtail its grand strategy.

Venice's Geostrategy

The geopolitical reality from the eleventh to the sixteenth century gave Venice an advantage over its competitors by placing it in a pivotal spot, mediating between the European and Asian centers of resources.[34] But it was not sufficient to guarantee Venetian power. Venice had to implement a foreign policy that reflected the underlying geopolitics. This meant extending control over sea arteries and the eastern Mediterranean outlets of Asian wares while limiting the expansion of its land border in Italy.[35]

Geopolitics did not determine the foreign policy undertaken by the Venetian doges. In fact, the decision in the fifteenth century to pursue a more proactive foreign policy on the Italian mainland, considerably expanding Venice's territorial base, was partly motivated by domestic interests and was not grounded in a sound reading of geopolitics. The consequence of this decision was that Venice weakened and lost its geographic advantage of being securely distant from Italian politics, serving as a powerful reminder that a grand strategy is effective only when it reflects the underlying geostrategic map.

Venice's grand strategy had three distinct moments in the years 1000–1500. From 1000 to 1204 Venice directed its efforts eastward in order to secure the increasingly vital avenues of commerce. This entailed four geostrategic steps: control over the Gulf of Venice, the conquest of Dalmatia, safeguarding Corfu and the entrance to the Adriatic, and finally the extension of its control over Constantinople and the eastern Mediterranean.

From 1204 to the late fifteenth century Venice had to manage the network of bases and ports upon which its power was grounded. This grid needed constant rearrangement to reflect shifts in Asian trade routes and changes in political conditions in the eastern Mediterranean. A system of bases and alliances imparts strategic leverage only insofar as it reflects the underlying map.

The decline of Venice started at the end of the fifteenth century, when Venice was at the peak of its political clout in Europe. Contrary to prevailing notions, the rising Ottoman challenge did not result in a drastic imperial retreat by Venice. Rather, the discovery of new routes to Asia shifted the power center of Europe toward the Atlantic, making the Mediterranean Sea a backwater region and Venice an insignificant power. Furthermore, Venice's pursuit of territorial gains in Italy altered significantly the geographic features of the republic. Venice became a land power with territorial commitments, facing the opposition of European states and Italian cities. As a result of the discovery of new routes and the expansion of Venetian power in Italy, Venice was no longer the intermediary between the European and Asian centers of resources and turned into a modest political actor in Europe.

The Rise of Venice

Venice was situated between two worlds: Byzantium to the east and Latin Europe to the west and north. As early as the ninth century this position allowed Venice to be an important commercial center through which the bulk of trade between Carolingian Europe and Byzantine and Arab ports passed. As an illustration of its role as a commercial broker, Venice accepted both silver and gold coins, used respectively by European and by Byzantine and Arab merchants.

Nonetheless, Venice's mediating location does not alone explain why it extended its influence over half of the Mediterranean. Venice had commercial and political interests in the two main centers of resources, Europe and the eastern Mediterranean, and could have expanded in either direction.[36] The rapidly developing markets in northern Italy and central and northern Europe were geographically closer to Venice, which was keenly interested in keeping them stable and open. However, until the fifteenth century Venice meddled only occasionally and diplomatically in Italian and European politics. Its main thrust of power was eastward.

There are two reasons for Venice's active interest in the East. First, the benefits of expanding eastward were greater than those of territorial gains in the Venetian "near abroad." An expansion in Italy would have gone against the interests of the papacy and the Western emperor, while the Byzantine emperor, facing the decay of his state, was initially eager to accept Venice's commercial and military influence. After Byzantium became too weak to offer political and commercial advantages to Venice, the Republic of St. Mark continued to be interested in the East. The desire to tap directly the Asian sources of wealth, for which Byzantium was only an intermediary, was a powerful incentive to expand eastward.

Second, on an operational level, in order to keep in contact with the two centers of resources Venice had to secure the necessary routes. On its western flank the routes to the Italian and European markets were manned safely and reliably by merchants coming from every corner of the continent. Venetians did not have to deliver the wares to their final destination in Flanders or Germany and often were happy to sell them to middlemen in commercial centers in Italy (such as Ferrara)[37] or France (at various fairs in Champagne). A network of trade and routes was already safely in place in Europe, and Venice had no interest in entering into a competition with the well-established European commercial actors.

To the east, however, there were no reliable merchant powers that could bring the goods from Byzantium, Alexandria, or the Black Sea ports to Venice and Europe. By the tenth century the Mediterranean power of Byzantium was in sharp decline, leaving the routes connecting it with Europe wide open to the plundering activities of pirates. The Adriatic and the eastern Mediterranean became a vacuum of sorts, severing the connection between Europe and Byzantium. As a result, to maintain its connection with the eastern markets Venice had to build its own network of friendly ports, alliances, and military bases.

Initially Byzantium, more preoccupied with the defense of its long Asian frontiers, supported Venice's efforts to establish a safe trade route in the Mediterranean. Between 1000 and 1204 the eastward expansion of Venice gradually filled the growing vacuum left by a retreating Byzantine power. As noted previously, Venice's expansion started in the surrounding lagoons, also known as the Gulf of Venice. Then, La Serenissima conquered Dalmatia and secured the mouth of the Adriatic. During the first two centuries Byzantine and Venetian interests coincided, but the beginning of the thirteenth century was marked by a clash of interests.[38] The final step in the expansion of Venice, in fact, was directed against Byzantium, which fell under Venetian control after the 1204 Fourth Crusade.

THE GULF OF VENICE

The first step in Venice's eastward expansion was the extension of her influence over the Gulf of Venice, the body of water between Venice and the Istrian peninsula. By the end of the tenth century Byzantium had withdrawn from most of Italy, and its presence in the northern Adriatic and on its coasts, the farthest region under its control, was dwindling. The main Byzantine port, Ravenna, decayed, leaving Venice as the last outpost of the Eastern Empire in this region. A strange relationship between Byzantium and Venice began: Byzantium needed Venice as an ally to patrol the Adriatic and bring its goods to the European markets, but at the same time Venice was a protectorate of the Eastern Empire. Since the late ninth century Byzantium had subcontracted part of its trade and, above all, the security of its western maritime periphery to Venice.

Venice expanded its influence relentlessly and strategically. Despite being a Byzantine subject, Venice always acted in its best interests, defending the Byzantine empire, while trading with its enemies. Beginning in the eighth century there were signs of a consistent Venetian presence in the Levant and in ports such as Muslim Alexandria, where in 829 Venetian merchants stole the body of St. Mark. In 992 Venice signed the first official treaty with Constantinople, by which it was granted trading privileges in Greek ports, lower tariffs, and special courts for its citizens. In exchange Venice had to help Byzantium defend its possessions in southern Italy, then under Saracen pressure.[39] The treaty of 992 had enormous significance. It indicated that while Byzantium was still in control of southern Italy and the southern Adriatic, it needed Venetian assistance to prop up its empire. Venice was thus free to extend its own protection over the northern Adriatic, where Constantinople was unable to keep order. Finally, by granting Venice generous trading benefits in the East, the treaty gave Venice a clear advantage over its commercial competitors (such as Amalfi). La Serenissima gradually loosened the imperial ties, becoming a de facto Byzantine ally.

By the end of the tenth century Venice was the most powerful city in the northern Adriatic. Its influence, however, was self-contained. By imposing economic sanctions, in 933 Venice forced Istria to sign a humiliating treaty, extending to this region its naval protection (in exchange for one hundred jars of wine per year) to prevent Slavic pirates from disrupting its commerce in natural resources with the Italian mainland.[40] Nonetheless, Venice's possessions were limited to the lagoon and the immediately adjacent mainland used as an outlet for the goods arriving in its port, and the Republic was careful not to extend its direct control over new territories.[41]

DALMATIA

The second step in Venice's rise to empire was the conquest of the Dalmatian coast. In 1000 Venice, under the leadership of Doge Pietro Orseolo II, put together a large fleet that took control of Dalmatia. The military expedition was the last stage of a delicate diplomatic process that maintained Venice under Byzantine influence while fostering courteous relations with the Western Empire. Both the Eastern and the Western Empire saw in Venice an ally that could effectively extend their dominion over Dalmatia and safeguard the commercial lifeline between Asia and Europe. Neither great power opposed Venice's expansion in Dalmatia, and the Adriatic Sea became increasingly identified as a Venetian sphere of influence.

As a result of Pietro Orseolo II's expedition, a string of cities, including the important port of Zara and the area around the mouth of the Naretva River, claimed allegiance to Venice. Although the Byzantine emperor was still mentioned in the oath of allegiance, the Venetian doge became duke of Dalmatia. Venice was again wary of establishing direct control over the Dalmatian cities, preferring to accept a simple promise of loyalty. This reluctance to add territories to the republic characterized much of Venice's history and empire. Venice's wealthy but small population preferred to focus on distant commercial ventures rather than on the administration of new territorial possessions.

Dalmatia was the necessary springboard to any further expansion in the Mediterranean. The abundant timber and the navigational importance of Dalmatia bestowed upon Venice control over shipping in the Adriatic and significant advantage in strategic resources. Influence carried responsibilities. The Dalmatian coast was constantly threatened by Slavic tribes from the interior and by Hungary, which stirred revolts against the Venetians throughout the sixteenth century. Within a few decades of the initial conquest in AD 1000 Venice was forced to send expeditionary forces several times to bring Dalmatia back under its control. However, the loss of a port (such as Ragusa) was usually offset by the extension of control over another city or island (such as Corfu). As a result, despite these recurring threats, from the beginning of the eleventh until Venice's decline in the sixteenth century, Venetian supremacy in the Adriatic remained unchallenged.

THE STRAITS IN THE ADRIATIC

A look at a map of the Adriatic shows that the Dalmatian coast was important for controlling the Mediterranean routes only as long as lower Dalmatia (modern Albania) and the heel of the Italian peninsula (Apulia, modern Puglia) remained under the control of a friendly power. The straits between these two coasts had a strategic importance similar to that of the Bosphorus, which controlled the entry to the Black Sea.

For Venice, as long as its patron, Byzantium, maintained its presence in southern Italy, the Adriatic remained an open sea. However, toward the end of the eleventh century Constantinople was in domestic turmoil, while the Seljuk Turks were pushing against its southern borders. Moreover, the 1054 schism between Rome and Byzantium further weakened the Eastern Empire because it clearly set it in opposition to the Roman papacy and to the Western Empire. At the Council of Melfi in 1059 the pope, unable and unwilling to rely on Byzantium for his defense, allied himself with the Normans, recognizing their dominion over Sicily, Calabria, and most importantly Apulia.[42] The Normans expanded rapidly, taking over southern Italy. In 1071 the key Adriatic port of Bari fell under the control of the Norman Robert Guiscard, thereby ending Byzantine rule of mainland Italy. In 1081 Guiscard crossed the straits between Apulia and the Albanian shore (between Durazzo and Corfu) and laid siege to the Byzantine city of Durazzo, on the Albanian coast, a key port that was also the western end of the old Roman Via Egnatia connecting Constantinople with the Adriatic. The Normans were ready to close the Adriatic, shutting Venice off from the Mediterranean and separating Byzantium from its European markets.

Venice came to the aid of Durazzo, respecting the 992 agreement with Byzantium. It also followed its interest by defending the gate to the Mediterranean Sea.[43] In a spectacular battle in June 1081 a Venetian fleet destroyed the Norman naval force, relieving Durazzo. Despite the defeat, the Normans maintained their presence at the mouth of the Adriatic for several years, until another naval defeat (Butrinto in 1085) and the death of Robert Guiscard the same year forced them to retreat.

For Venice the lower Adriatic was a key strategic objective. Without control over it Venice could not project power to the eastern Mediterranean. In fact, to achieve control over the straits between Otranto and Apulia, Venice had to devote all of its resources, while other Italian powers, such as Genoa and Pisa, were reaping benefits from the Crusades in Syria and Palestine. A renewed Norman threat in the 1140s, accompanied by Hungarian attacks against Dalmatia, again forced Venice to maintain the bulk of its naval forces in the Adriatic, missing the territorial and commercial gains obtained by Genoa and Pisa from the first crusades.

The lasting Norman menace compelled Byzantium to maintain friendly relations with Venice and grant it increasingly larger benefits. After Guiscard's expedition to Durazzo was defeated, Byzantium rewarded Venice with ample commercial privileges. The 1082 bull of the Eastern emperor granted free trade rights to Venice and gave the doge full jurisdiction over Venetian merchants throughout the empire, including the city of Constantinople. After the second attempt by the Normans to block the mouth of the Adriatic, Venice forced the Byzantine emperor in 1147 to renew and widen the privileges of the 1082 bull. Venetian quarters in Constantinople were enlarged, and Doge Enrico Dandolo was granted the aristocratic title of *protosebastos* in perpetuity. By becoming the military guarantor of the Byzantine Empire, Venice also became the principal trading power in the Mediterranean.

AGAINST BYZANTIUM

The growing power of Venice strained its relations with Byzantium. Venice became increasingly assertive of its independence from the Byzantine emperor. For instance, starting in the mid-twelfth century the elected doges made a *promissio ducalis,* a promise to the Venetian people, disregarding the oath to the Eastern emperor.[44] Moreover, the interests of the rising but formally subordinate republic began to diverge from those of the decaying but still imperial Byzantium, and tensions flared. For example, during their joint expedition against the Normans in Corfu (1147) the Venetians and the Greeks started fighting between themselves, resulting in a failed attempt to retake the island. During the following years the Venetians did not hesitate to raid Byzantine merchant vessels in the Aegean whenever the opportunity presented itself.

The Eastern emperor was wary of the growing Venetian influence. Several times he attempted to balance Venice's privileged position in the empire by giving similar benefits to its Pisan competitors. In the early twelfth century, however, Pisa was entangled in a conflict with Genoa and failed to take advantage of the Byzantine offers. Moreover, Byzantium needed Venice more than Venice needed imperial privileges. Byzantium was, in fact, clearly dependent on the Venetian naval power to maintain control over its imperial structure in the Mediterranean. A breaking point occurred in 1171, when Emperor Manuel Comnenus arrested all the Venetians in Constantinople. This was an act of desperation rather than one of deliberate policy, reflecting Byzantium's weakening state as compared with the power and arrogance exhibited by Venice. In response, Venice readied a powerful fleet of one hundred galleys and twenty transport ships, which sailed toward Constantinople in the fall of the same year. After bringing Ragusa back under Venetian control, the fleet anchored in Negroponte. There the Venetian expedition was struck by plague, caused, according to some sources, by poisoned wells. Instead of quarantining the fleet, Doge Vitale Michiel sailed back to the lagoon, bringing with him the plague. The enraged population killed him on the step of a convent where he was seeking refuge.

Venetian diplomacy continued to work and accomplished what a military expedition failed to achieve. Venice cut its ties with Byzantium and worked effectively against it. In 1175 Venice signed a treaty with the Normans normalizing relations and, consequently, leaving Byzantium without a naval defender. Moreover, after the 1176 battle of Legnano, in which Frederick Barbarossa was defeated by the Lombard League (which included Venice), ending his plans to rebuild an empire in Italy, Venice brokered a peace between the emperor and the pope. The Peace of Venice had an important impact on the geopolitical arena of the Mediterranean: Venice was recognized as a great power and, above all, as the principal Mediterranean naval power. The Western emperor allowed Venetian merchants to import goods from anywhere in the empire, while his own subjects could travel only "as far as the city and no farther, a restriction that seems to recognize the Adriatic from the lagoon southward as Venetian waters."[45] Such recognition increased Venice's power and leverage over Byzantium, leaving the Eastern Empire isolated in the face of a united political front in Europe. In 1183 the Eastern emperor freed the Venetian hostages held in Constantinople since 1171, renewing trading privileges and restoring Venice's position in the eastern Mediterranean.[46]

The split between Byzantium and Venice was finalized with the Fourth Crusade in 1204. The objective of the crusade, initiated by the pope in France, was to reconquer the Holy Land. In order to reach this objective, however, the Christian knights and barons needed a fleet that could transport them to Syria or Egypt. Venice offered its naval services and in 1201 signed a contract with the French barons that stipulated the transport of 20,000 foot soldiers, 4,500 knights, 9,000 squires, and 4,500 horses for the price of 85,000 marks.[47] When the crusaders arrived in Venice, it became clear that they were unable to pay the agreed sum. In an event that continues to be hotly debated by historians, the cunning Doge Enrico Dandolo compelled the crusaders to give Venice half of the lands expected to be conquered, while keeping the destination of the expedition vague. Moreover, Dandolo convinced the French knights to help the Venetian fleet to conquer Zara, on the Dalmatian coast, on their way to the Holy Land. Zara was a strategic port because it served as a key supply base for ships headed to Constantinople and the Egyptian and Syrian ports, as well as a significant source of oak, which was indispensable for shipbuilding and difficult to obtain in Italy.[48] During the 1180s Zara revolted against Venice and put itself under the protection of Hungary and, later, Pisa. In the first military engagement of the Fourth Crusade Venice put down Zara's rebellion and, with the help of the German and French crusaders, retook control of that port.

The crusade was diverted toward Constantinople after the conquest of Zara. Whether this diversion was the result of Venice's premeditated plan is one of the most contested issues in medieval history. One side argues that the Venetians, with Doge Dandolo as their leader, betrayed the Christian goodwill and fervor by hijacking the crusade to achieve commercial benefits for their city. In light of the profits the Venetians obtained after the crusade, this argument appears quite plausible.[49] The other side, however, defends Venice by extending the blame to the rest of the crusaders, the French and the Germans, who were eager to pillage the wealthy capital of the Eastern Empire, partly in revenge for the lack of Byzantine support for the previous crusades. During the early crusades, in fact, the Eastern Empire had been too weak and too close to the Muslim-controlled territories to engage in a continuous holy war against them, preferring to seek a form of coexistence.[50] Finally, some Venetian apologists argue that the diversion of the crusade was a result of "a series of mistakes, failures, and accidents" that had been unforeseeable.[51]

The truth is probably somewhere in between. The Venetian leaders, with Enrico Dandolo as the doge, had a keen appreciation of the strategic importance of Constantinople for the political and commercial success of their city. Enrico Dandolo in particular, a figure of Periclean authority and power, was a sharp strategic thinker who was the "mind of the [Venetian] body politic."[52] He was aware that Venice was geostrategically disadvantaged at the end of the twelfth century. During the previous crusades Venice, preoccupied with Norman invasions and Byzantine vacillations, had played a secondary role. In the early twelfth century, in exchange for commercial benefits, it had offered only modest naval help in the conquest of the port of Tyre and the cities of Haifa and Acheron. Pisa and Genoa, on the other hand, had been actively involved in transporting crusaders to Palestine, obtaining duty-free ports and markets, which gave them an advantage in trading with the Muslims in Egypt and in Syria.[53] In particular, in the last decade of the twelfth century Pisa was a serious challenge to Venice, attempting several times to close the Adriatic. Therefore, participating in the Fourth Crusade was a way for Venice to regain lost strategic ground. The crusade presented an unparalleled occasion to extend Venetian influence over the eastern Mediterranean with the aid of French and other European forces.

Furthermore, despite the 1183 restoration of peaceful relations with Byzantium, Venice continued to nourish imperial desires toward the Aegean and the eastern Mediterranean. The 1187 fall of Jerusalem made the alternative routes through the Aegean, the Bosphorus, and the Black Sea all the more important. Consequently, the city of Constantinople assumed greater strategic importance: the power that controlled the city had access to the key Black Sea route. Given the habit of the Eastern emperor to try to balance Venetian influence by dispensing benefits to Pisan and Genoan merchants, Constantinople was not firmly on Venice's side. Peace with Byzantium was not beneficial for Venice.

It is clear that Dandolo and his advisers were aware of the enormous strategic gains that an expedition against Byzantium could bring.[54] This is not to say, however, that the Venetians concocted the whole expedition in order to strike directly at Byzantium. Contingency played an enormous role. The immediate cause of the detour of the crusade was the appearance among the crusaders of the "young Alexius," son of the dethroned emperor Isaac, who convinced the leading barons to restore him to power in Constantinople. In exchange he promised them commercial benefits, as well as money to pay their outstanding debt to the Venetians. Greed was stronger than the religious zeal of the crusaders, who, in any case, justified their assault against Constantinople by arguing that the Eastern Empire had rejected the pontifical authority of Rome and consequently was heretical. The Venetians could not help but be glad of this fortuitous turn of events.

In the summer of 1203 young Alexius and his father, Isaac, were brought back to Byzantium by the Western fleet. The crusaders obtained the promised commercial benefits but did not continue their expedition to the Holy Land, remaining near the city, on the shores of the Golden Horn. The protection of the newly acquired benefits in Constantinople carried greater appeal than the holy struggle against the Muslims. The crusaders' decision to remain near the city was not made in vain. In February 1204 Alexius was strangled by one of his courtiers, who upon becoming the new emperor, Alexius V, rejected the claims of the Western crusaders. The Venetians convinced the rest of the Western force to conquer Constantinople again. In April the crusaders entered the city for the second time, pillaging it for three days.[55] After the wanton destruction, they elected their own emperor, Baldwin, Count of Flanders.[56] Venice, assisted by French and German soldiers, had vanquished its former patron.

In theory the crusaders established a "Latin empire" on the ruins of the Byzantine Empire. In practice the new Latin empire existed only on a map, and from the very start it was a hollow political entity. The Byzantines had not been completely subjugated and set up a parallel structure of power in a city only fifty miles from Constantinople. Demoralized because of their inability to defeat the resisting Byzantine force, less than a year after the second conquest of Constantinople several thousand crusaders, the bulk of the Western force, deserted, thereby weakening the military foundation of the Latin empire. Moreover, the crusaders were deeply divided by inner conflicts exacerbated by the inflow of knights and crusaders from Palestine in search of quick and easy territorial gains. For the Venetians, a big setback was the death of the ninety-seven-year-old Enrico Dandolo in 1205, an event that led to further confusion among the Latin crusaders.

Venice never took possession of its three-eighths of the Byzantine Empire. Only by title the doge was the "Lord of One Quarter and One Half of a Quarter of the Empire." The Venetians, however, were not interested in the extent but in the location of territory. They developed a structure of bases that, although territorially small, gave Venice a controlling position in the eastern Mediterranean. As agreed upon at the outset of the crusade, the Venetians were granted three-eighths of the city of Constantinople and developed a powerful merchant colony that even commanded its own galley fleet. To strengthen their control over the northern access to the Byzantine capital, they also extended their protection to the city of Adrianople.

Paradoxically, despite the initial interest in Constantinople, that city did not become the pivot of Venice's power.[57] Immediately following the crusade, some Venetians proposed to move the seat of the Republic of St. Mark to Constantinople. However, the debates were inconclusive because the Venetians preferred to run their commercial interests from the safety of the lagoons. The Adriatic was by itself an important source of wealth and would have lost its centrality had Venice moved to Constantinople.[58] Moreover, Constantinople attracted the attention of other powers, including Venice's main competitor, Genoa. The new Latin emperor's promise to exclude the Genoese from all Greek ports, including Constantinople, proved to be too difficult to implement. Genoa harassed Venetian trade wherever it could. In order to limit the damage and to concentrate on the administration of the newly acquired empire, in 1218 Venice signed a treaty allowing Genoese ships to enter Constantinople.[59] After the end of the Latin empire in 1261, the Genoese replaced Venice in the capital of a resurrected Byzantine Empire.

The focus of Venetian geostrategy lay in the Adriatic and the Aegean. In the Adriatic, Venice obtained Zara during the Fourth Crusade and Corfu in 1206. Genoese pirates controlled this strategic island, disappearing in the face of the approaching crusading fleet but returning soon thereafter. Only after the conquest of Constantinople did a Venetian fleet return to Corfu and defeat the pirates. Venice did not hold the island for long, giving it up soon thereafter. Nonetheless, the Venetians maintained control over the Adriatic and the link between their naval bases through two ports in the Ionian Sea, Coron and Modon, on the southern tip of Morea. Ships entering the Adriatic were required to make a call at these ports, which became the "two eyes of the Republic."[60] Together with Zara and Ragusa on the Dalmatian coast, Modon and Coron were part of the structure of bases that gave Venice a dominant position in the Adriatic.

In the Aegean, Venice gained several strategically located islands. The jewel of the new Venetian empire was Crete, with Candia as its main port. After the 1204 conquest of Constantinople, Crete became the principal objective of Venetian military efforts, opposed by Genoa. In the end, Venice conquered the island, establishing over it a feudal rule. Through Crete Venice controlled access to Syria, Egypt, and the Aegean routes to Constantinople and the Black Sea. Crete allowed Venice to dominate the eastern Mediterranean.[61]

In Negroponte Venice obtained trading rights and a port, consolidating its chain of bases in the Aegean. Furthermore, through a series of private expeditions, led by two of Enrico Dandolo's nephews in search of glory and wealth after the crusade, Venice extended its control over the Cyclades (with the fertile Parox and Naxos wrestled from the hands of Genoese pirates) and the island of Andros. As a result, the Aegean was under Venetian domination, although less so than the Adriatic.[62] The geography of the Aegean made it difficult for one power to have control over it. While the Adriatic could be controlled by shutting the straits between Otranto and Corfu, the Aegean was an open sea. To control the Aegean, Venice had to extend its military authority over several islands.

The Fourth Crusade was the last step in the construction of the Venetian empire. Venice rose from a Byzantine subject to a great power that replaced the Eastern Empire as the largest maritime power in the Mediterranean. Having implemented a geostrategy that mirrored the geopolitical reality, Venice now faced the challenge of keeping it relevant.

The Management of the Empire

In the decades immediately following the Fourth Crusade Venice was at the peak of its power. La Serenissima controlled the Adriatic, the Aegean, and Constantinople and was one of the most powerful states in Europe. However, Venice did not remain at the peak of its power for long, for three reasons. First, during the Middle Ages and the Renaissance technology limited a state's ability to exercise exclusive control of the sea. Second, Venice faced a new challenger, Genoa, that posed challenges to the management of its empire. Third, and arguably most importantly, Venice became increasingly more entangled in adventures on the Italian mainland.

During the Middle Ages and the Renaissance control of the sea meant something very different from our modern concept. At best, control of the sea was a seasonal reality. In calm weather ships could stay out of their bases for longer periods of time. During winter most ships were confined to ports and rarely ventured on long seafaring journeys. Consequently, during the winter months the sea was a no man's land, leaving space to state-sponsored or private pirate activities. At the beginning of spring Venice had to refit a galley fleet to clear the sea of pirates and make it safe for the spring commercial trips.

Even during the summer, control of the sea depended on the control over key bases, ports, and small bodies of water. Galleys, the principal ships used for military operations, required frequent resupplying of food and water for their large crews of oarsmen. As a result, they were unable to patrol the sea for prolonged periods of time and required a closely spaced chain of bases. Throughout the fifteenth century the Venetians preferred the galley to other, less labor-intensive ships because it could be navigated on windless days, was easier to maneuver in ports and in military formations, and, thanks to its large crew, was a formidable vessel for naval battles and amphibious attacks.

Thus, control of the sea meant control of strategic bases. The strength of a power, in this case Venice, was based less on the size and capability of its fleet than on the extent and cohesion of its network of ports and bases. By controlling key bases Venice was able to deny access and navigation to its challengers. Conversely, a challenger could weaken Venice's grip on trade with the Levant by depriving it of a few pivotal ports. Naval battles, therefore, had strategic meaning only insofar as they were fought over ports or key islands.[63] As the 1571 battle of Lepanto would show, the destruction of a state's fleet did not signify the end of its naval power.

Genoa posed a challenge to Venice's string of bases, thereby threatening the foundations of Venetian power. Already after the Fourth Crusade it was evident that Genoa was Venice's main antagonist. Most of the islands and ports that Venice conquered in the eastern Mediterranean had to be wrestled from Genoese, not Byzantine, hands. With the rapid end of the Latin empire, Genoa restored some of its strategic advantage by helping the new Byzantine emperor, Michel Paleologue, to regain his throne in Constantinople in 1261. A series of naval battles between Venice and Genoa followed: near Salonica in 1262, in Settepozzi in 1263, and off Trapani in 1266.[64] The Republic of St. Mark won most of the battles, but Genoese pirates continued to harass Venetian cargoes, imposing serious costs on Venice.[65]

The resurrection of the Byzantine Empire weakened Venice. In 1270 France brokered a peace, but the war between Venice and Genoa continued until a Venetian defeat in a battle near the island of Curzola in 1298. A new peace treaty was signed on surprisingly equal terms: Venice recognized Genoese primacy on the Ligurian coast, while Genoa agreed not to enter the Adriatic unless its ships were headed for Venice.

The uneasy balance of power between Venice and Genoa subsisted for half a century. Neither power was able to inflict a mortal wound on the other; they preferred to pillage each other's trade. Venice had a geostrategic advantage because it dominated not only the Adriatic, making it a key artery for Italian and European trade, but also the Aegean and Crete, giving it control over commerce with Egypt and Syria. Genoa was at an advantage in Constantinople and consequently in the Black Sea trade, but even there Venice was able to extend its influence.[66] Despite the animosity aroused by the establishment of the Latin empire, the Venetians were willing to negotiate with the new Byzantine emperor, and Genoa had no choice but to accept Venice's return to the Bosphorus and the Black Sea. Moreover, the rising Ottoman threat in the eastern Mediterranean made Venice, the most powerful naval power, also the natural bulwark of Europe. As a result, European powers, such as a rising Spain and France, favored Venice in its struggle with Genoa.

In 1350 a new war with Genoa started, but it was a war between two powers weakened by the plague and domestic tension. A few years earlier the Black Death had decimated the populations of the two cities, diminishing their armies and fleets. Venice was able to muster the support of both Byzantium and Catalonia, who were eager to limit Genoese ambitions in their respective regions. Even with the contribution of Catalan and Byzantine forces, the Venetian fleet numbered only about one hundred galleys, half the size of the 1204 fleet.

Initially Venice had the upper hand, winning a battle near its base of Negroponte. Later on, however, Genoa won in the Bosphorus, after a long battle fought during the night and in the midst of a storm, and near Modon. The adverse military results were mitigated by skillful Venetian diplomacy and by Genoa's domestic unrest. Genoa was forced to accept Milan's rule to quell civil strife. Moreover, Milan was eager to end tensions with Venice, which had important mainland allies, including Holy Roman Emperor Charles IV, threatening Milanese territories.

Venice ended the war weakened but still holding a decisive geostrategic advantage. During the war Venice had lost Dalmatia to the king of Hungary, but it maintained exclusive control over the Adriatic by preventing Hungary from developing its own fleet. Moreover, despite its losses in the Aegean and Constantinople, Venice continued to control the bulk of European trade with India. In fact, political upheaval in Persia redirected the main Asian trade routes away from the Black Sea and Syria to the Red Sea and Egypt. In other words, Genoa's control over the Bosphorus and the Black Sea was not as lucrative as before because the underlying map had changed.

Changes in the geopolitical map, however, are difficult to read. As a result, states often continue to fight for locations that are of little geostrategic significance. For example, Venice continued to pursue the Black Sea trade, again antagonizing Genoa. In the 1370s it conquered Tenedos, a small island strategically located at the mouth of the Dardanelles. In one of the first episodes of the so-called gunpowder revolution, Venice installed cannons on the island to defend it and to increase its ability to control access to the Bosphorus.[67]

Venetian actions sparked the fourth war with Genoa, the War of Chioggia (1378–81). Responding to its loss of Tenedos, Genoa attacked Venice where nobody else had dared to attack: in the lagoons. With the help of the king of Hungary in Dalmatia, Genoese ships reached the Venetian lagoons in 1379 and conquered Chioggia, on the southern shore. The city of Venice, however, remained free, and the Venetians counterattacked, expelling the Genoese from Chioggia, the lagoons, and finally the Adriatic. With the 1381 Treaty of Turin a victorious but seriously weakened Venice gave up fortifications on Tenedos as well as in Dalmatia. Moreover, it was forced to grant trading rights to Genoa in Cyprus, weakening its hold over the maritime routes in the eastern Mediterranean.

Finally, the War of Chioggia proved to many in Venice the fragility of their power, which was based exclusively on access to the Adriatic and Mediterranean sources of food and wealth. Genoa's foray into the Adriatic had threatened the survival of Venice because it had not only attacked the imperial center, the city of Venice, but also severed it from the only source of its power, the maritime trade routes. The lesson some leaders in Venice learned from the war with Genoa was that Venice needed to take a greater interest in the Italian *terraferma*.[68]

The Decline

Genoa's actions did not cause the demise of Venice. Nor, contrary to many historical interpretations, did the rise of the Ottoman Empire. In fact, paradoxically, the rise of the Ottoman Empire helped Venice to strengthen its string of bases in the Aegean and the Adriatic, leaving Genoa even weaker. The Greek cities, fearful of Muslim advances and aware of Byzantine fragility, eagerly turned to Venice for protection. At the end of the fourteenth century La Serenissima acquired Durazzo in Albania, Lepanto, several ports in Morea, more land in Negroponte, and a few islands in the Aegean. By the beginning of the fifteenth century Venice had returned to Dalmatia and restored its rule on the Adriatic coast.

The relations between Venice and the Ottomans were too complicated to be classified as antagonistic. Commercial interests brought Venice in close contact with the Ottoman Empire. Despite recurrent pontifical edicts prohibiting trade with the Muslims, Venetians merchants were spread all over the Ottoman territories, as well as in Muslim Egypt. As a result, Venice was reluctant to engage openly in war with the Ottomans and preferred an accommodating policy, even though this generated hostility from other European powers that were fearful of the Ottoman threat. It is significant that despite the involvement of some Venetian individuals, Venice did not come to the rescue of Constantinople in 1453. The Byzantine capital fell into the hands of the Ottomans, to the great disconcertion of the Western world but to the apparent nonchalance of Venice.[69]

Despite the accommodating policy of Venice, a clash between increasingly powerful Ottomans and the extensive Venetian empire was unavoidable. The Venetian network of bases and ports intruded on Ottoman space and on the direction of Ottoman conquests. After the 1453 fall of Constantinople the Ottomans continued their westward expansion, striking the Venetian possessions in the Aegean. In 1470, after a siege during which the Muslims skillfully used cannons to keep the Venetian fleet at bay, Muhammed II the Conqueror himself led the conquest of Negroponte, ending Venice's control over this important island. As after the 1261 loss of Constantinople, the extensive imperial framework of Venice compensated for the loss of Negroponte, maintaining Venetian influence over the Aegean and the trade with Egypt and consequently Asia. The Ottomans did not continue their expansion in the Aegean, and in 1479 they signed a peace treaty with Venice that, although clearly in their favor, left the influence of the Republic of St. Mark in the eastern Mediterranean almost intact. As Frederic Lane observed, the Ottomans never "closed the doors to Venetian traders."[70]

It was only at the end of the sixteenth century that the Ottoman power fatally weakened the Venetian system of bases in the Mediterranean. In 1571, despite Europe's naval victory at Lepanto that same year, Venice lost the important base of Cyprus to the Ottomans. But by then Venice was already a hollow power. The spectacular victory achieved at Lepanto was more a sign of Venetian weakness than of Venetian strength. Venice's fleet, in fact, had to be buttressed by significant Spanish naval forces, which led the battle. Venice's losses in the face of the Ottoman Empire were a result, not a source, of its decline.[71]

The decline of Venice began in the West, not in the East. Crucial to Venice's slow retirement from the world scene were the gradual expansion of La Serenissima into the Italian mainland, absorbing increasingly larger forces and attention, and a dramatic change in the geopolitical reality caused by the discovery of new routes to the Indies and the resulting shift of the seat of European power toward the Atlantic.

EXPANSION ON THE *TERRAFERMA*

Venice had always had a close relationship with the Italian mainland for two reasons: food and trade. The Venetian lagoons had never been able to produce food and fresh water in sufficient quantities to feed the population, and from its early days Venice had had to secure the supply of these goods by controlling a limited number of possessions on the mainland. The example of the water supply is illustrative. Through ingenious innovations, such as artificial wells called "Venetian wells," which collected rain, Venice mitigated the problems resulting from its location in the middle of a salty lagoon.[72] When Venice's population reached one hundred thousand in the fourteenth century "Venetian" and artesian wells could not produce enough water, and it became necessary to import water from the mainland. For this purpose the Republic of St. Mark began to draw fresh water from nearby rivers, such as the Brenta. Barges would fill up with water from upstream to guarantee its freshness and deliver it to wells in Venice. The importance of the Brenta River was underscored in 1142, when the Venetians went to war with the Paduans because they were threatening to divert its course.[73]

A second reason for Venice's involvement on the Italian mainland was the need to maintain stable trade routes linking the city (and its maritime empire) with the markets in Europe. This meant protecting the rivers Po and Adige, through which ships loaded with Eastern wares reached internal markets in Lombardy, from where trade routes continued to northern Europe. Moreover, Venice obtained most of its food from the fertile lands of the Po valley, increasing the importance of mainland Italy.

The doges had traditionally been cautious in intervening on the mainland, limiting themselves to sporadic sorties to restore the balance of power and to protect access to rivers. For instance, during the invasion of Frederick Barbarossa in the second half of the twelfth century Venice intervened on the side of the Lombard League, composed of the majority of northern Italian cities, with the purpose of preventing the extension of German hegemony to this region. After Barbarossa was defeated at Legnano in 1176 and the peace treaty was signed in Venice, La Serenissima retreated to the safety of its lagoons and to the expansion of its overseas bases.

Venice was more proactive in defending access to rivers. For example, in 1240 Venice fought a war with Ferrara to secure control over the mouth of Po. Half a century later, in 1308, the Venetians tried to put Ferrara under their control, irking the pope in the process. Despite the pontifical excommunication, the Venetians defeated Ferrara by threatening to build a canal to the Adige through which the majority of their trade would reach the Lombard markets, making the Po less important. Finally, in 1329 Venice conquered several cities, such as Verona and Padua, after they tried to tap Venetian wealth by levying heavy duties on goods being shipped to and from the lagoons.

Until the fifteenth century, interventions by Venice in northern Italy were only occasional. There was no grand plan to conquer the region and add a sizeable land component to the empire. The Venetians were averse to the acquisition of territory, feeling more comfortable at sea and controlling few small but strategic possessions.

In the late fourteenth century this traditional strategy was slowly abandoned in favor of a policy aimed at building a strong and direct Venetian control over the Italian *terraferma*. The War of Chioggia alerted Venetian leaders to the importance of having a solid base on the Italian mainland, considered to be indispensable and complementary to the well-developed maritime empire.[74] Moreover, in the early fifteenth century the growing power of Milan, ruled by the ambitious Gianga-leazzo Visconti, altered the map of Italy. By gradually incorporating several cities in northern Italy Visconti ended the age of medieval localism. A unified power in northern Italy presented a threat to all the other cities and states on the peninsula. Initially, to the disappointment of Florence, Venice was blind to the risks of Milanese hegemony and befriended the Viscontis to gain a few territories around Treviso, on the path to a few important Alpine passes.[75] When Milan vacillated after the death of its leader, Giangaleazzo Visconti (1402), Venice conquered Vicenza, Verona, and Padua, slaughtering the ruling Carrara family.[76] The con-quest of this power vacuum was relatively easy, spurring Niccolò Machiavelli to observe that "Florence [had] spent more on wars and acquired less than Venice."[77]

However, the ease with which Venice conquered a large part of northern Italy was a mixed blessing. In fact, the rapid and violent extension of Venetian power to these territories deprived Venice of the strategic frontier offered by the lagoons. The lagoons ceased to be a bulwark against attacks. Moreover, by extending its territory westward Venice occupied the string of cities that were a useful buffer between it and Milan. By 1410 Venice's territories bordered directly on the state of Milan, which, under a new Visconti leader, continued to be the most powerful and aggressive state on the Italian peninsula. The conflict between Florence and Milan ended with the Peace of Lodi (1454), which established a league of Italian cities, including Venice, aimed at preserving Italian independence. However, the league did not survive its signatories (Cosimo de Medici in Florence and Fran-cesco Sforza in Milan), and by the end of the fifteenth century Italy was again divided by strife. Venice maintained neutrality, if not a friendly attitude, toward the Viscontis and did not oppose Milan.

The consolidation of Milanese power began to preoccupy Venice only after Genoa, weakened by the War of Chioggia and torn by civil strife, was included in Visconti's state in the second decade of the fifteenth century. Italy's map suddenly became dangerous for Venice. Three land powers—Milan, Florence, and Naples—were vying for hegemony over Italy. If the Duchy of Milan and the Kingdom of Naples succeeded in conquering the city of Florence, Venice would be left alone, facing a unified power on the *terraferma*.

The Venetians were divided as to the direction of their geostrategy. Some ar-gued for a withdrawal from Italy, while others insisted on the necessity to main-tain a strong Venetian presences on the mainland. An early-fifteenth-century debate between the two camps serves as a powerful illustration of these two conflicting strategic visions. In 1421–23 Venice had to respond to a diplomatic opening by Florence, which was appealing for Venice's help. The debate that followed, considered to be a focal moment in Venetian history, resulted in the end of the traditional policy of isolation or offshore balancing. One camp, led by Doge Tommaso Mocenigo in the twilight of his life, argued that the strength of La Serenissima was in the maritime empire in the East. Italy, he argued, was and should be securely separated from Venice by the lagoons. Moreover, "the territory of the Visconti was so essential for Venice's prosperity, by providing food and raw-materials and by using a large share of Venetian trade, that Venice must never allow a war to be waged against Milan and devastate the 'garden' of Venice."[78] A war between Venice and Milan was a war between a seller and a buyer. Neither could gain much from it. Mocenigo's policy recommendation was to keep Venice neutral, avoiding the temptation of further territorial expansions. A retreat into the lagoons was no longer feasible, but the hilly and defensible region of Verona should be the farthest Venetian stronghold on the *terraferma*. Florence could balance Milan by itself. In brief, Mocenigo advocated the continuation of the age-old Venetian policy of strategic aloofness from Italian politics.

The other camp, led by the young Francesco Foscari, was pro-Florentine and consequently in favor of Venetian involvement in Italian politics. Foscari asserted the indivisibility of security: what harmed Florence also harmed Venice. The argument was heavily influenced by ideological overtones that defined interests on the basis of domestic similarities, not of geostrategic realities. Florence was a fraternal republic fighting against the encroaching tyranny of the Viscontis. Therefore, it was in Venice's interests, as well as Venice's obligation, to join forces with the Florentines because the preservation of liberty abroad was the first step in the preservation of liberty at home.[79]

Foscari was elected to the dogeship. Venice followed his policy and, as a Floren-tine ally, entered into a war with Milan. Venice abandoned the role of an offshore balancer and became deeply enmeshed in Italian politics. The resulting struggle with Milan and the defense of its Italian territories absorbed much of Venice's energy. Because of their small population and lack of skills in land warfare, the Venetians were obliged to hire several condottieri, such as the expensive Gonzaga of Mantua, Francesco Sforza, and Bartolomeo Coleoni. These mercenaries not only commanded high salaries but also had an inherent interest in prolonging a conflict between states.[80] Some, such as Francesco Bussone, known as Carmag-nola, conspired against their employers in favor of Milan.[81] Moreover, Venice had to divert large quantities of military materiel to the land struggle with Milan, at one point sending more ships upstream on the Po than to patrol the Aegean. For instance, in a 1438 battle to relieve Brescia the Venetians transported by land two galleys and eighty other ships to Lake Garda in an impressive but wasteful show of force. Two years later they built a fleet of eight galleys directly on the shores of Lake Garda, gaining control over it.[82]

By the mid-fifteenth century a large portion of northern Italy had been acquired by Venice, and the Republic of St. Mark had become a land power in Italy. The Venetians "were returning to the mainland, not as bargemen peddling salt and Eastern fabrics, but as rulers directing fleets and armies."[83] The financial and military costs associated with this land expansion were large but paled in comparison with the political damage.[84] In fact, after the defeat of Milan Venice became an object of jealousy and fear, and its offensive in northern Italy created alarm in the rest of Europe. As Riccardo Fubini observed, "Venice trespassed upon imperial rights and was so indifferent to agreements and loyalty among princes that the myth of its unlimited expansionist desires spread even beyond the Alps."[85]

In 1466 Galeazzo Sforza, Duke of Milan, wrote to the Venetians:

> You do a grevious [sic] wrong, you possess the fairest State in Italy, yet you are not satisfied. You disturb the peace and covet the states of others. If you knew the ill-will universally felt towards you, the very hair of your head would stand on end. Do you think the states of Italy are leagued against you out of love to each other? No; necessity has driven them. They have bound themselves together for the fear they have of you and of your power. They will not rest till they have clipped your wings.[86]

In fact, Florence turned against Venice, its fellow republic, because it feared that Venice could have conquered Milan and then Italy. Cosimo de' Medici decided not to support Venice and to favor Milan instead. "Had he not done so," Francesco Guicciardini writes, "the Venetians without doubt would have become lords of that state [Milan] and shortly thereafter of all Italy. In this case, the freedom of Florence and of all Italy was safeguarded by Cosimo de' Medici."[87]

Venice did nothing to assuage Italian fears. After a brief interlude of Italian unity against Carl VIII's French troops in 1494, Venice continued to pursue its territorial ambitions in Italy. In fact, in 1499 the Venetians themselves invited the new French king, Louis XII, to Italy to help them conquer Milan. It is widely accepted that this decision was an enormous political blunder.[88] The alliance between France and Venice was unnecessary; it was fueled by Venice's territorial ambitions, which had no geostrategic justification. Machiavelli argued that this alliance brought ruin to Venice because La Serenissima was the weaker power and because after the alliance won, "the weaker prince remain[ed] prisoner" of the stronger.[89] From then on, foreign powers, such as Spain and France, would continue to descend upon Italy to tilt the balance of power in favor of their Italian allies, gaining their own territories in the process.

The two main goals of Italian powers were to keep foreign invaders out and to keep Venice in check.[90] By the end of the fifteenth century Venice was threatening both goals. It had brought the French power to northern Italy, while continuing to expand. After the failed attempt of Cesare Borgia to restore pontifical power over central Italy, Venice directed its power southward toward the papal territories in Romagna. To counterbalance Venice's power, in 1509 the pope and the Holy Roman Emperor formed the League of Cambrai. The King of Hungary, the Duke of Savoy, the king of France, the king of Spain, and several small Italian cities joined the league, eager to grab the spoils of Venice. The Republic of St. Mark was defeated on the *terraferma* at the battle of Agnadello (14 May 1509), losing its possessions in Lombardy.[91]

When the French became too powerful, Venice skillfully rearranged the opposing coalition in its favor and regained some of the lost territories.[92] Nevertheless, it was too little too late. The 1529 the Peace of Cambrai hindered Venetian expansion in northern and central Italy. Moreover, it heralded the advent of Spain as the new power in Italy and in the Mediterranean. Venice became a second-rank power that was contained and balanced on the Italian mainland while no longer dominating in the Mediterranean Sea.[93] From then on, as in the case of the 1571 battle of Lepanto, Venice would wage war against the Ottomans only when supported by Spain. The best it could do was to ally itself with a stronger power.

Venice mangled its grand strategy in Italy. It lacked strategic vision and became entangled in a "war of ambition."[94] As Machiavelli observed, "Few in Venice could see the danger, even fewer could see the remedy, and nobody could advise it."[95] By becoming involved on the Italian mainland Venice lost its geostrategic detachment from the deadly balance-of-power politics. Its power was wasted in futile territorial conquests.[96] The devastating War of Cambrai showed to Venice that "a seafaring nation, which aspired to rank as a land power also, and relied on mercenaries for its defense, was throwing away the secret of its strength."[97]

NEW DISCOVERIES AND CHANGE IN TRADE ROUTES

The discovery of new routes to the Indies contributed to the decline of Venice. The voyage of Columbus and particularly that of Vasco de Gama fundamentally changed the geopolitical map faced by Venice. The discovery and exploitation of

new routes to the Asian markets first by the Portuguese and then by the Spaniards, Dutch, and English transferred the seat of European power from the Mediterranean to the Atlantic. As a result, Venice lost the role of Europe's "entrepôt."

The Venetians immediately understood the dramatic change caused by Vasco
de Gama's voyage around Africa. The news of his trip arrived in Venice in 1499,
after a year's delay.[98] The first rumors of such a trip reached the Republic of St.
Mark through Egypt, whose Muslim leaders probably were afraid of losing the
profitable commerce with Venice.[99] When Venetian embassies in Spain and Portugal confirmed the news, the Monte Nuovo, Venice's stock market, lost 50 percent of its value in one day.[100] Venetian merchants were aware that they would
lose their monopoly in the trade of Asian spices because of the possibly lower
prices offered in Lisbon. Moreover, if the Portuguese tapped Asian goods directly
in India, they would cut Venice off from the source of this lucrative trade. As a
result, Venice would lose power and perish, as Girolamo Priuli put it, like "a child
deprived of milk and food."[101]

Venice's response was quick but desperate. To undermine the commercial rise
of its challenger, Venice refused to purchase spices in Lisbon. Venetians enjoyed
a monopoly in the spice trade in ports such as Alexandria and would not have
been able to have a similar standing in Lisbon. Portuguese merchants, unlike
their Muslim counterparts in Alexandria, did not need Venice to deliver their
wares to the European markets in Germany or France. They developed their own
trade with those markets, and the Venetians would have been an unnecessary
intermediary.

Most importantly, despite the fundamental change in the map of the world,
Venice continued to defend the geostrategic status quo. In the first years of the
sixteenth century Venetian diplomacy strengthened Venice's relationship with
the sultan in Egypt. Venice even encouraged the sultan to stop Portuguese supplies at their source, in India.[102] Moreover, to expand the Mediterranean access
to the Asian markets and make the arduous Persian land routes obsolete, Venice
on several occasions (1504, 1530, 1586) encouraged Egypt to build a Suez canal.[103] In 1586 excavation work had even begun, but in the end nothing came of
these projects.

In Venice the Portuguese competition drove the spice supply down and prices
up. For much of the first decade of the sixteenth century Venice was forced to
abandon its trade with Alexandria because it was no longer profitable. However,
higher prices were not the only cause of the decline of Venice's commercial
power. In the 1560s Venice regained some of the lost trade by lowering prices.

After the Portuguese had to fight the Ottomans in Asia and were incorporated into the Spanish Empire, their élan weakened. The war waged by Spain in Holland further depleted Lisbon's resources. Nonetheless, the deviation of Portuguese efforts away from Asia was momentary. The conquest of supply sources in Asia by Portuguese, Spanish, and, later, Dutch and English merchants undermined Mediterranean trade. By destroying Arab naval forces, Portugal succeeded in disrupting the spice trade between India and the Red Sea.[104]

Venice had always had difficulties in gaining direct access to the Asian markets. Marco Polo's voyages to China in the late thirteenth century were spectacular but commercially ineffective. Throughout the centuries Venice tapped Asian markets through intermediaries in Persia, Egypt, and the Central Asian steppes. The Atlantic powers of Europe, in contrast, reached Asia directly, circumventing the Middle East. Moreover, they built a network of ports and colonies (such as Portugal's Calicut and Goa) that consolidated their discoveries.[105] Venice's trade continued until well into the eighteenth century. In fact, in absolute terms it continued to grow. But the circumnavigation of Africa and later the development of the American colonies marginalized Venetian power, which in relative terms became increasingly smaller.

Cut off from the sources of its trade, the Mediterranean became a regional sea. The bulk of European trade moved to the oceans, where Venice could not compete.[106] In fact, as noted elsewhere, Venetian ships were not outfitted for long sea voyages, especially in rough oceanic waters. The Atlantic caravel was more ocean-worthy than the Mediterranean galley.[107] Stuck in the relative safety of the Adriatic and the Mediterranean, Venice failed to modernize its naval force and continued to live in an earlier technological era.[108] Thus, a secure geostrategic position can thwart the advancement of military technology, allowing the state to continue its political life relatively undisturbed without modernizing its military.

Furthermore, Venice's geostrategy and system of bases reflected a geopolitical reality that no longer existed. Its bases in the Adriatic and the eastern Mediterranean were situated on trade routes that had become increasingly irrelevant. Venice had no capability and perhaps no strategic vision to develop a network of bases and routes that could counterbalance the one rapidly built by the Atlantic powers. The idea of building a Suez canal seems to indicate that Venetian leaders attempted to expand their geostrategic framework outside of the Mediterranean in order to protect their supply sources. However, the idea failed to materialize, in part because of technological deficiencies, leaving the Venetian empire at best as a regional actor in Italy.

Conclusion

The historian Jacob Burckhardt observed that "Venice recognized itself from the first as a strange and mysterious creation—the fruit of a higher power than human ingenuity."[109] Indeed, the rise and the decline of Venice were owing in large degree to the "mysterious" influence of geography. However, "human ingenuity" in the form of grand strategy played an important role. The ability of Venice to conduct a foreign policy reflective of the underlying geopolitical map constituted its main strength. As John Godfrey observed, "A reason for the Venetian Republic's success was that more than any other State it understood the importance of geography in politics and commerce. The Venetians were to develop an unerring eye for good harbours, trade routes, and strategic islands. They always knew just how much of an area it was necessary to occupy, and they were not interested in conquest for its own sake."[110]

When Venice's foreign policy failed to reflect the underlying geopolitical reality, the Venetian empire was doomed to decline. Part of the divergence between Venice's grand strategy and the geopolitical reality was caused by a misguided policy pursued by the republic in Italy. The resulting alteration of Venice's borders was crippling, and it could have been avoided. However, part of this divergence between policy and reality was owing to a radical change in the underlying geopolitical map caused by the discoveries of new routes and new lands. Given Venice's geographic location and system of bases and allies, it would have been difficult, if at all possible, to adapt to the new map. Geography, therefore, played a crucial role in the history of Venice. As Venice's "geographical isolation had determined her rise to empire, so geographical considerations foredoomed her to decay."[111]

From this study of the Venetian empire we can draw three conclusions. First, the geopolitical reality, understood as a map that changes with new discoveries, displacements of trade routes owing to technological advances, and shifts in the locale of power, is a powerful variable that can influence the fate of a state. In the case of Venice, geopolitical changes had a determining impact on Venetian power, making its empire strategically irrelevant.

Furthermore, the ability to formulate and implement a foreign policy that reflects the underlying geopolitical reality is the key to achieving and maintaining a position of power, if not supremacy. A state's geostrategy must be flexible enough to respond to changes. The less flexible it is, the greater the dependence of the state on the immutability of the geopolitical situation. The value of Venice's network of bases and alliances was conditional on the immutability of trade routes from Asia. When the routes changed, the Venetian framework lost its strategic relevance.

Finally, in order to maintain an empire it is necessary to avoid premature geostrategic changes (e.g., through retrenchment or further expansion). A grand strategy advocating the abandonment of traditional allies or bases because of fatigue or lack of domestic interest, and not because of fundamental changes in the geopolitical map, undermines the strength of a state. Such a change can be damaging not only to the imperial framework but also to the very independence of the state in question. In fact, states are more than their homeland. They are, in a sense, global in nature. Their economic welfare and political power are often dependent on regions that are distant. In the fifteenth century Venice forgot the importance of its distant bases and was lured to an excessive degree into land conflicts in Italy. The strengthening of its presence in the "near abroad" did not lead to greater power. On the contrary, it led to the growing enmity of the great powers of Europe and the eventual decline of Venice.

CHAPTER FIVE

The Geostrategy of the Ottoman Empire (1300–1699)

On 12 September 1683, at six o'clock in the afternoon, a heavy cavalry detachment of twenty thousand German and Polish soldiers charged the Ottoman army from the hills the north of Vienna. It was the largest cavalry charge until then. Under the command of the Polish king, Jan Sobieski, the cavalry broke through the Ottoman lines and, together with the soldiers led by the French duke Charles Lorraine, penetrated the Ottoman camp. In the succeeding hours the European allies completely defeated the Ottoman army. The Ottoman vezir, Kara Mustafa, barely escaped the battlefield "on one horse and with one change of cloths," leaving behind him 15,000 dead Ottoman soldiers, 117 cannons, several tons of gunpowder, a lavishly equipped encampment, and his parrot.[1] The Ottoman siege of Vienna, and with it the imperial élan of the Ottoman Empire, was over.

Why were the Ottomans in Vienna? Why did their defeat near this city start their decline? These questions beget a broad question of what Ottoman geostrategy was and where the Ottomans directed their attention and power. Like the coeval Venetian empire, the Ottoman state responded to an underlying geopolitical situation defined by trade routes and centers of resources, and pursued a geostrategy that attempted to mirror this reality. That is, the Ottoman sultans sought access to resources and control over strategic routes, while being vigilant on their land frontiers. And the empire prospered when its geostrategy matched the underlying geopolitical reality. When the geopolitical situation changed with the discovery of new routes to Asia, and Ottoman geostrategy failed to adapt (or, as in the case of the battle of Vienna or the failed expedition to India, was prevented to expand toward strategic regions), the empire began to decline.

Ottoman geostrategy was characterized by a relentless focus on Europe. Geographically, at its peak the Ottoman Empire stretched from the outskirts of Vienna to Baghdad and Algiers, and from the time of Muhammed II (1451–81) the

Ottoman sultan wore the title "Ruler of Two Continents and Two Seas."[2] Yet the title of the sultan, as well as the geographic extent of his empire, is misleading because from the first decades of the fourteenth century Ottoman geostrategy focused mostly on the European continent. The Middle East (Persia, Egypt) and North Africa and the Mediterranean and Red seas were peripheral for the Ottoman Empire. They were relatively poor, and after the fifteenth century they were no longer located on the main strategic routes linking Europe and Asia. In the mid-sixteenth century the Ottomans tried to restore the commercial link to Asia through the Middle East. Failing, they continued to focus on central Europe and the Danube valley.

The Ottomans' pursuit of control over the wealth and routes of central Europe continued despite a powerful, albeit inconsistent, European opposition. If states want to maintain or increase their power, they have to expand toward resources and routes, regardless of the presence of a countervailing power. But this meant that the Ottoman Empire was almost continually at war with Europe, which despite dramatic early defeats continued to oppose Ottoman advances. When the Europeans succeeded in checking the Ottoman imperial push in Vienna and along the Danube River, the sultans and their state began a slow and inexorable decline. After the 1683 battle of Vienna, the gradual but relentless loss of their European territories deprived the Ottomans of resource-rich regions and of control over strategic routes in central Europe, increasing simultaneously the instability on their northern land frontiers.

Why the Ottoman Empire?

Foreign-policy scholars have avoided studying the rise of the Ottoman Empire. Instead, the lengthy decline of the Ottoman Empire, spanning two centuries up to World War I, spawned prolific studies of international relations. The Ottoman Empire was the "sick man of Europe," propped up by foreign powers (Great Britain, the Austro-Hungarian Empire, Russia, and France) who feared a power vacuum in the middle of Eurasia and astride the vital Bosphorus and Dardanelles straits.[3] But these studies focused on the great powers rather than on the Ottoman Empire and often lacked an in-depth understanding of the origins of this empire and of the motivations underlying its foreign policy until the eighteenth century.

The studies of the rise of the Ottoman Empire are also seriously limited. Most examine only the establishment of Ottoman dominance in Anatolia and across the straits. Often clouded by legends and a lack of primary documents, the main puzzle of these studies is the rise of the Ottoman tribe to the leading position in

Asia Minor.[4] Classic studies, such as those by Paul Wittek and Herbert Adams Gibbons, begin with the legendary figure of the first Ottoman sultan, Osman, in the thirteenth century. Most of them end with the Ottoman conquest of Constantinople in 1453, which consolidated the Ottoman state as a European great power.

In this chapter I build on the latter category of studies, focusing on the foreign policy of the Ottoman Empire up to and past the 1453 fall of Constantinople. I argue that the rise of the Ottoman Empire and the direction of its expansion have been strongly informed by the geographic distribution of power and access routes in Eurasia. The Ottomans initially succeeded because they pursued a foreign policy that mirrored the geopolitics of that period. My argument complements and does not contradict the most accepted interpretations of Ottoman history. Some historians, such as Wittek, stress the religious characteristics of the rise of the Ottomans, while others, such as Gibbons, stress the Ottomans' tribal strength. However, these historians rarely examine in detail the role of geography in the rise of the Ottoman Empire.[5] My goal is to expand on this neglected subject by examining the impact of geopolitics on the Ottoman Empire.

The Ottoman case offers important lessons on the impact of geopolitics on geostrategy. Specifically, the history of the Ottoman state raises the intriguing question why it did not become a sea power, or why it directed its efforts to territorial aggrandizement in Europe rather than maritime control, unlike other Mediterranean powers such as Venice or even Byzantium. The Ottomans, in fact, built a state that was concerned predominantly with its land borders and whose main expansion was on land. And despite the fears of other Mediterranean and European powers, the Ottoman Empire never became a naval power equal to Spain, Venice, Portugal, or, later, Great Britain. While the empire succeeded in spreading terror to the Italian and Spanish coastal communities, the sultan could never claim to rule the Mediterranean Sea or most importantly, from the early sixteenth century on, the Atlantic and Indian oceans.

In part the geopolitical situation of the Ottoman Empire explains why the sultans did not focus on the sea. The most important centers of resources were located to the north of Asia Minor, in Europe, and to the far east, in Asia. The wealth of these two regions was a prize coveted by Mediterranean powers (Genoa, Venice, and the Ottomans), and the wealth in Asia was also coveted by the rising Atlantic states after 1500 (Portugal, Spain, the Netherlands, and, after the eighteenth century, Great Britain). In order to increase their relative power the Ottomans had to expand toward Europe and, after the arrival of the Portuguese in the Indian Ocean, Asia in order to control resources and routes.

At the same time, the Ottomans were constrained in their pursuit of control over resources and routes by their continental location. Despite audacious forays on the Mediterranean and Red seas, the Ottoman state was severely hampered by its long and unstable land frontiers. As A. T. Mahan affirmed, "If a nation be so situated that it is neither forced to defend itself by land nor induced to seek extension of its territory by way of the land, it has, by the very unity of its aim directed upon the sea, an advantage as compared with a people one of whose boundaries is continental."[6] For a sea power the loss of a distant base does not constitute a major geostrategic setback, especially if the lost outpost can be circumnavigated or replaced by another port or a different trade route. For a land power every threat or attack against a land border constitutes a direct menace to its core. Therefore, the principal geostrategic concern of a land power is its borders, which are easier to threaten and cross than coastal frontiers. The longer and the more unstable the land frontiers, the less attention and resources the state can devote to distant, noncontiguous areas. Consequently, as a land power, the Ottomans were at a disadvantage; forced to devote their energies and attention to their European and Asian borders, they had to relegate their maritime ambitions to the realm of pretensions.

A further reason why the Ottoman Empire failed to become an important naval power was that by the time it had consolidated its territory and extended its control over the Mediterranean shores, this sea had become geopolitically passé. By the mid-sixteenth century the Atlantic powers, spearheaded by the Portuguese, were masters of the Atlantic and Indian oceans. Although they did not completely cut off the supply of Asian wares to the Mediterranean, they established a powerful alternative to this traditional route, undermining its strategic importance. Because of its geographic position and particularly of its continental borders, the Ottoman Empire remained locked at best in the Mediterranean, at worst in the territories on both sides of the Bosphorus. The situation of the Ottoman Empire in the 1500s illustrates well the rigidity of the geostrategy of a continental power. The land borders limit the choices available to a land power, restricting its ability to redirect its geostrategy in response to geopolitical changes.

Alternative Explanations

Before developing my argument in greater historical detail, it is appropriate to deal with two alternative explanations of Ottoman geostrategy. The first suggests that Ottomans were motivated by religious zeal and expanded toward territories populated by Christians. Their rise as well as their decline is directly correlated to their adherence to this religious vision. The second asserts that Ottomans ex-

panded where there was a power vacuum and that their geostrategy was successful when it was directed toward regions with no counterpoising power.

The first explanation is the most prevalent in the literature. Paul Wittek asserts
that the reason for the rise and strength of the Ottoman Empire was the "ghazi
factor," an almost fanatical impulse to proselytize. Ghazis were the Muslim warriors who fought to expand Islam, and a *ghaza* was a Muslim holy war. According
to Wittek's argument, in the twelfth and thirteenth centuries the Ottomans, pre-
senting themselves as conquerors of Christian territories, had a powerful attraction for neighboring Turkish tribes. Nomad soldiers from as far as the Caucasus
flocked to the small Anatolian region under Ottoman control, seduced by promises of booty and, above all, of a good fight against the Christians living on the
other side of the Bosphorus. The perception of being an active force of conversion
to Islam was a powerful factor in establishing authority over Muslim populations.[7] Consequently, the expansionistic nature of the Ottoman state was, according to Wittek, a result of the need to be perceived as the warriors of Islam, as
ghazis. "From the first appearance of the Ottomans, the principal factor in this
political tradition was the struggle against their Christian neighbors, and this
struggle never ceased to be of vital importance to the Ottoman Empire."[8] The
decline of the Ottoman Empire was connected to the weakening of the proselytizing fervor. When the Ottomans clearly abandoned the *ghaza* by allying themselves with the Catholic Habsburg Empire during World War I, their empire
collapsed.

The ghazi-factor explanation suggests the Eurocentric nature of Ottoman geostrategy. As long as the Ottomans fought against Christian Europe, they could
attract manpower and authority. However, this argument should not be exaggerated. While it is true that some sultans, such as Osman or Bayezid I, were religious fanatics, they did not establish a state geared to converting "infidels."[9]
In fact, the Ottomans rarely forced conversions to Islam. Their conquests in
Europe proved to be stable and lasting because, among other reasons, the Ottoman administration allowed freedom of religion. With few exceptions the Ottoman sultans, despite being the leaders of the supposed *ghaza* against Christians,
were very tolerant and often surrounded themselves with non-Muslims. During
Muhammed the Conqueror's reign all foreigners, regardless of their religion,
were protected legally. In the case of the Jewish population, according to Louis
Thuasne, "it can be said that it enjoyed a greater freedom than the Jews living
in the Christian countries."[10] For instance, in 1479 Muhammed the Conqueror
chose a Jew to carry a sensitive letter to the Venetian government.[11]

Often the Janissaries, the elite military unit created by forcefully converting young Christian children taken from European territories, is given as an example of the ghazi factor.[12] But this elite force was not meant to act as an ideological force. The Janissaries were trained to fight, not to convert.[13] It is true that the Janissary corps was a product of the Ottoman plan to expand in Christian Europe, not in Muslim Asia. But it was a result of operational, not religious, needs. In Asia Minor an army of quick horsemen was more than enough to subdue the local tribal warlords, who rarely commanded vast fortifications. The walls and forts of Constantinople and of the Western powers in the Danube valley required a different army. Horsemen were "not well suited for siege operations against fortified towns of the Byzantine Empire."[14] The establishment of the Janissary corps, therefore, marks the shift from nomadic warfare, based on speed, mobility, archery, and horses, to a heavier force whose aim was the conquest of Europe.[15]

Finally, the *ghaza* explanation does not fully account for the other directions of Ottoman expansion. For instance, Ottoman campaigns at the end of the thirteenth century and the beginning of the fourteenth were against fellow Muslim Turkish tribes in Anatolia. The subjugation of Anatolian princes was a key objective in the early Ottoman plans because they presented a threat to the rising Ottoman power in northern Asia Minor.[16] Even Wittek admits that the Ottomans directed their effort toward Christian Europe only when the Anatolian regions were either too strong to be conquered or stable enough not to require Ottoman military interventions.[17] Contrary to the *ghaza* theory, the Ottoman Empire fought wars against Muslim tribes and populations on its southern border throughout its history. It expanded in Persia, Syria, Egypt, and North Africa to the detriment of Muslim, not Christian, populations. In particular, Ottoman authorities were highly skeptical of mystical and fanatical Muslim sects, such as the Shiites in Persia, which were considered a serious threat to Ottoman imperial control of the Middle East.[18]

In conclusion, the *ghaza*, or holy war, argument cannot fully explain the Ottoman geostrategy. While the enmity between Muslim and Christian populations played some role in attracting soldiers to the Ottoman ranks and in directing the Ottoman geostrategy toward Europe, it was not the main factor shaping Ottoman foreign policy. It was Europe that led crusades against the Ottoman Empire, not vice versa. The Ottomans pursued a geostrategy with a keen eye on their geopolitical situation. Theirs was not an ideologically motivated holy war but a geopolitically informed grand strategy.

The second alternative explanation of Ottoman geostrategy is the neorealist "power vacuum" thesis.[19] According to this argument, states respond to the distribution of power in the world, expanding where they find no counterpoising power. Geostrategy, therefore, is shaped by the assumption that power deters and weakness attracts.

This alternative explanation asserts that Ottoman geostrategy was directed toward weaker neighboring regions. For instance, according to this argument the Ottoman push against the Byzantine Empire, culminating in the 1453 conquest of Constantinople, was provoked by the chronic weakness of the latter, which simply attracted the attention of the sultans. Similarly, this argument asserts that the southward drive of the Ottomans in the sixteenth century was motivated by the political and military fragility that characterized Persia and Egypt. The lack of an opposing force determined the direction of Ottoman geostrategy.

My argument contradicts this alternative explanation. States expand toward resources and access routes, regardless of the presence of a counterbalancing power. It would be naive and dangerous to assume that the geostrategy of states is based only on the evaluation of the disposition of power and that they expand only toward weaker regions or "power vacuums." In fact, states tend to expand in search of resources and routes and do not respond only to the fear of clashing with other powers. States that control resources and strategic routes wield enormous influence, and as European powers experienced in their relations with the Ottoman Empire, it is often difficult to deter them from expanding toward these sources of power.

The Ottomans expanded mainly toward Europe because of the importance of its resources and trade routes in central Europe. The presence of powerful counterbalancing states did not deter the Ottomans. Indeed, after the collapse of Byzantium the Ottoman Empire faced the greatest European powers and did not desist expanding along the Danube valley toward the heart of Europe. Contrary to the "power vacuum" hypothesis, the primary geostrategic objective of the sultans was also the most powerful neighbor of the Ottoman Empire.

The Rise of the Ottoman Empire

The geopolitical situation of the Ottomans, and their correct reading of it, was crucial to their rise to political prominence in Anatolia and then in Europe. The Ottoman tribe, in fact, was situated close to the European center of resources, in particular the fertile lands in the Balkans and in the Danube valley. At the same time, it sat astride vital trade routes connecting Asia with the growing European economies. The Ottomans took advantage of this geopolitical situation and expanded in three directions: north to Europe, south to the Middle East and the Indian Ocean, and southwest to the Mediterranean. By heeding the geopolitical

Map 2. The three directions of Ottoman geostrategy

situation the Ottomans became the unquestioned regional leader, controlling southeastern Europe, the Middle East, and North Africa.

I examine the period from the legendary birth of the Ottoman state in the thirteenth century to the end of Ottoman expansion in the seventeenth century. After the 1683 battle of Vienna, in Ranke's words, "the Ottomans ceased to be feared, and began themselves to fear."[20] From then on, Ottoman foreign policy was driven mainly by survival concerns, losing its expansionistic vigor. The Ottoman state from the thirteenth century to the end of the seventeenth century, therefore, exemplifies a great power seeking to expand in clear geographic directions, in this case reflecting the underlying geopolitical situation. Its success as well as its gradual failure illustrate the constraints imposed by geopolitics upon a state's geostrategy.

The Rise of the Ottoman Tribe: Toward Europe

The rise of the Ottomans was helped by the geography of Anatolia in two ways. First, by encouraging social dispersion the geography of Asia Minor hindered the creation of alliances that could have counterbalanced the rise of one tribe. The inaccessible internal valleys, separated from coastal areas, made communications difficult. As a result, the different Turkish tribes had few incentives to unite, create alliances, and settle. "Geography made for disruption, and explains the strong tendency in Anatolian history toward the formation of small social and political entities," wrote William Langer and Robert Blake.[21] The resulting political plurality of the region facilitated the rise of the Ottomans, as there was no counterbalancing coalition among the other tribes.

Second, within Anatolia the geographic location of the Ottoman tribe was fortunate. The Ottomans occupied an area in the northern part of Anatolia, close to the fertile coastal strip on the Sea of Marmara and the Dardanelles and Bosphorus straits. The easy access to these regions, as well as to the even wealthier ones across the straits in Europe, gave the Ottomans a strong advantage over their neighboring tribes in Anatolia.[22] In fact, because of their proximity to European wealth the Ottomans became the most "attractive" tribe in Asia Minor.[23] "Nomad soldiers streamed to the leaders who could promise them the 'best wars,' wars against settled Christian kingdoms that would provide the most booty."[24] Moreover, in the thirteenth century a large flow of population from Central Asia and the Caucasus to Asia Minor, caused by Mongol invasions, injected new strength into the region occupied by the Ottomans.[25]

From the very beginning of the Ottoman state's ascendancy its strength depended on its capacity to expand in Europe. As the *ghaza* explanation posits, the faster the Ottomans expanded in Europe, the greater was their attraction for the other tribes and people.[26] But there was more to this than religious zeal. The wealth of European regions gave the Ottomans a material advantage over the poorer nomadic tribes in southern Anatolia. As a result, the geographic foundation of the Ottoman Empire was in Europe, not Asia.

The political situation in Europe facilitated Ottoman expansion. For one thing, the European political scene was fragmented and unable to offer a coherent response to the initial Ottoman push. In the thirteenth and fourteenth centuries the main European powers were preoccupied with the most immediate threats to their security. The rising powers France, Great Britain, and Spain were devoting their strategic attention to one another and, later on, to the Atlantic. Venice, the most directly affected by the ascent of a potential challenger in the eastern Mediterranean, was entangled in a long struggle with Genoa in the fourteenth century and in territorial conquests in Italy later on. The papacy in Rome, another likely opponent to the rise of a Muslim power, was in the midst of a crisis that culminated in the 1378 split between Rome and Avignon. Closer to the Ottomans, the Balkan Slav princes were divided, weak, and more preoccupied with wrestling power from Byzantium than with the nascent Ottoman Empire. The crumbling empire of Constantinople attracted more attention than the rapid rise of the Ottoman tribe. The result was that the big players of Europe were at best ignorant of, at worst indifferent to, the rapid Ottoman expansion. When in the fifteenth century they noticed with great alarm the Ottoman power, the sultans were already firmly entrenched in Europe, having vanquished the Byzantine Empire and conquered Constantinople.

The second, more immediate factor that helped and encouraged Ottoman expansion in Europe was the political situation of the Byzantine Empire. Since the eleventh century, the Eastern Roman Empire, which shared a border with the Ottoman state, had been in terminal decay. The weakness of Constantinople became clear at the turn of the thirteenth century, when the Byzantine capital fell under the assault of an expedition led by the Venetians in 1204. It never recovered its luster. For instance, more than 90 percent of the population of Constantinople had left the city by the early thirteenth century.[27] At the beginning of the fourteenth century, when the Ottomans were already one of the main Anatolian tribes, Byzantium was unable to garrison its Asian frontiers. Most of the Byzantine troops were needed to keep the restless Balkan princes under control, leaving Asia Minor wide open. The Eastern emperors simply could not fight a two-front land war. Facing a geopolitical situation similar to the one that would be faced by the Ottomans a few decades later, they had to make a strategic choice between defending (or expanding in) the European north or the Asian south. Constantinople decided in favor of its northern regions, oblivious to the rising Ottoman power in the south.

The weakness of Byzantium's Asian frontier, combined with the wealth of its territories, attracted the attention of Turkish tribes. These tribes, "driven by despair, lack of space, desire for booty and religious fanaticism, pressed on with astonishing force and number into the western provinces of Asia Minor, which by 1300 were hopelessly lost."[28] Furthermore, in an attempt to sway the domestic balance of power in their favor the various factions of the Byzantine Empire allied themselves with the nomadic tribes of Asia Minor, soliciting them to fight on

European shores. For example, in 1308 a group of Turks crossed the straits into Europe, invited by the notorious Catalan Grand Company, a band of mercenaries hired by pretenders to the Byzantine throne.[29]

These marauding interventions in Europe set the direction of Ottoman expansion. The Turkish soldiers who fought on Byzantine payroll, "returning from this expedition with rich spoil, made it clear to the Ottoman ruler in which direction lay his future conquests. The goal was Thrace and Macedonia, the way led over the Dardanelles and the peninsula of Gallipoli."[30] The Turkish tribes led by Osman, the founder of the Ottoman dynasty, slowly nibbled at the Byzantine frontier lands and cities. About 1326 they conquered the first Byzantine city, Bursa. Three years later the Ottomans entered Nicea, a city that was considered the "storehouse of Greek wealth in Asia Minor."[31] In 1337 Nicomedia, situated "on the sea at the head of a noble valley through which the great highway leads into the interior of Asia Minor," surrendered after a six-year Ottoman siege.[32] Bursa, Nicea, and Nicomedia gave the Ottomans fertile lands and, above all, access to the sea.

As mentioned earlier, the Ottomans crossed the straits into Europe, helped by the domestic factionalism of the Byzantine Empire. In the 1340s John Cantacuzenos, a pretender to the Byzantine throne, allied himself with the Ottomans in order to defeat the emperor John Palaelogos. To seal the alliance the sultan, Orkhan, married Cantacuzenos's daughter.[33] As a result of this alliance the Ottomans crossed the Sea of Marmara and laid their first, unsuccessful siege to Constantinople in 1346. Six years later Cantacuzenos again looked to the Ottomans for help against John Palaelogos. This time, in exchange for his military services the Ottoman sultan obtained a fortress in Thrace, the first Ottoman foothold in Europe. And although John Cantacuzenos was eventually defeated, depriving the sultan of a valuable ally, the Ottomans were already firmly established in Europe, and the new Byzantine emperor failed to evict them. As Gibbons wrote, "John Cantacuzenos introduced the Osmanlis into Europe. John Palaeologos accepted their presence in Thrace without a struggle."[34]

Once Ottoman armies crossed the Sea of Marmara, they quickly strengthened their position on the European shores. After a serendipitous earthquake in 1354 weakened the walls of Gallipoli, the Ottomans occupied the city, strategically located at the western entrance to the Sea of Marmara. This was probably the most important strategic conquest until the fall of Constantinople a century later.[35] It offered "prospects of limitless expansion towards the west."[36] As Inalcik observed, the conquest of Gallipoli "not only gave the Ottomans a strong position from which to establish themselves in Thrace, but also from which to control sea traffic between the Mediterranean and the Black Sea."[37]

After the conquest of Gallipoli the Ottomans expanded rapidly and relentlessly in Europe.[38] Until then Ottoman forays into Europe, led mostly by local warlords, had been for the purpose of plundering or assisting a Byzantine faction. Once the Ottomans had established an outpost in Europe, they kept invading across the straits with the conscious strategic goal of conquering new territory and solidifying a powerful presence in Europe.[39] In 1356 the Ottoman sultan led his troops in the conquest of two key Byzantine cities, Demotika and Adrianople, "the greatest Ottoman advance yet made in Europe."[40]

A few years later the sultan Murad continued the Ottoman expansion into the Balkan peninsula.[41] Following the Maritza River, he attacked the Bulgarian and Serb states. At the same time, in 1363 he strengthened Ottoman control over Adrianople and Philippopolis, severing the trade route connecting Constantinople with its main sources of grain and foodstuffs in the Balkans.[42] The Byzantine ruler, weak and alone, signed a treaty with the Ottomans, de facto recognizing them as an equal power and promising not to reconquer his lost territories in Thrace. Moreover, Byzantium agreed to join forces with the Ottomans against their common Anatolian enemies.[43] In 1365 Ottoman power was further recognized by the Adriatic city of Ragusa, which asked to be allowed to trade with the Ottomans in exchange for a yearly tribute.

Despite these Ottoman victories, European powers failed to unify and organize an expedition to check their rapid advance. A few isolated voices warned European powers of the impending Ottoman threat. For instance, in 1355 the respected Venetian leader Marino Faliero wrote to the Venetian Senate "that the Byzantine Empire must inevitably become the booty of the Osmanlis, and urged his countrymen to get ahead of them."[44] Faliero's warnings about the Ottomans remained unheeded, while his invitation to take over the Byzantine Empire was alluring.

In fact, instead of trying to save the Eastern Empire and using it as a bulwark against the rising Ottomans, the European powers wanted to weaken it even further. Byzantium was still considered to be a heretical city that did not recognize the authority of the Roman pope. The Crusades were spurred as much by fear of the Ottomans as by hatred of the Byzantines. Often the hatred of the Orthodox Church was greater than the fear of Islam, and European powers played the Ottomans against Constantinople. In the mid-1360s, for instance, Genoa helped Sultan Murad transport sixty thousand of his troops across the straits into Thrace to extract concessions from the Byzantine emperor.[45]

Given Europe's suspicious attitude toward Byzantium, it is not surprising that the crusades against the Ottomans failed.[46] For instance, spurred by the pope, an anti-Ottoman crusade was organized in 1363; three years later it succeeded in wresting control of Gallipoli from the Ottomans. But the crusaders failed to extend this strategic gain, instead directing their attention to the Bulgarians, on the Black Sea. The Ottomans were still considered to be an insignificant power unworthy of Europe's attention. After the 1366 crusade Pope Gregory XI attempted to organize a second crusade against the expanding Muslims. However, no Christian power was willing to contribute money or manpower to fight the Ottomans, while the pope himself was reluctant to defend Byzantium unless the Eastern emperor renounced his heretical stand.[47]

The lackluster European response allowed Sultan Murad to conquer the Balkans. Europe's weakness was compounded by disagreement among the Balkan kingdoms. For instance, the death of the Bulgarian czar Alexander caused deep internal divisions among the Balkan populations, eroding the opposition against the Ottomans. In 1366, just after the failed European crusade, the Ottomans defeated the Serbs. In 1371 the Serbs suffered another crushing defeat on the Maritza River, where their king was killed. The Ottomans extended their empire to Serbia, Macedonia, and Bulgaria. In 1373, in a gesture of unprecedented humility, the Byzantine emperor recognized the Ottoman sultan Murad as a sovereign, promising to render the sultan military service and consenting to surrender his son Manuel as a hostage.[48]

The European nature of the Ottoman Empire begets one of the first questions in this chapter: why did the Ottomans expand north, not south? During the fourteenth century, while the Ottoman sultan transferred his capital from the Asian shore to the European city of Adrianople, the land border in Anatolia remained almost unchanged. Ottoman geostrategy was clearly directed toward Europe, not Asia.

The two alternative explanations discussed above have some relevance here. The anti-European, hence anti-Christian nature of Ottoman geostrategy attracted a considerable number of Muslims in the early stage of expansion. Ottoman geostrategy was in fact directed toward the collapsing Byzantine Empire, which offered little, if any, resistance to the advancing Ottomans. For instance, many Byzantine cities fell under Ottoman control because they had been abandoned by the Byzantine army rather than because the Ottomans had developed an efficient army capable of assaulting a well-defended town. The Ottomans expanded by controlling the lands surrounding a city, depriving it of food and access to the rest of the Byzantine Empire. Left to themselves, the cities preferred to open their gates to the Turkish soldiers.

Yet, the reason for Ottoman expansion was not simply the lack of a counterpoising power or the religious zeal of Anatolian tribes. The Ottomans had to expand in Europe, instead of in the barren region to the south, in order to become the ruling power in Anatolia and in both Europe and the Middle East. According to Gibbons, "From Europe, Asia Minor and more could be conquered: from Asia, no portion of Europe could be conquered." In fact, he continues, Ottoman strategy in Asia was conditional on continued successes in Europe.[49]

The wealth and skills necessary to build an empire were in Europe, not in Asia. By expanding in Europe the Ottomans acquired fertile lands that supplied food and wood in great quantities, as well as abundant manpower. For instance, the Ottomans obtained their supplies of grain from the lower Danube region.[50] In southern Anatolia, in contrast, there was nothing that could have made the Ottomans a great power. The dry, isolated valleys were a large power vacuum, but they were difficult to control and led to no strategic trade route or commercial entrepôt. Important caravan routes connecting Asian markets with the Mediterranean passed through Persia and Syria, but it was more important to control the termini, such as Constantinople and other ports, than to control the routes themselves. To control these ports meant to control the flow of goods transported via the land routes. The Ottomans would not have gained influence and power by expanding into Anatolia and the Middle East.

Only after he had anchored Ottoman power in Europe did Sultan Murad direct his attention southward. The awe inspired by Ottoman exploits in Europe was so great among Anatolian tribes that they readily accepted the sultans' authority. Not until 1378 did Murad have to resort to arms, to take control of the city of Tekke, southwest of the lake region in Asia Minor. The Ottomans were able to extend their influence over Anatolia simply because of their conquests in Europe.

But Murad did not extend Ottoman influence beyond southern Anatolia; rather, he limited himself to subduing a few problematic tribes. In fact, a wider expansion in Asia would have been politically dangerous for the Ottomans. In the fourteenth and fifteenth centuries, European powers appeared to constitute the most menacing threat to the Ottomans. Had the sultans decided to embark on an expansion southward, Venice, Genoa, France, the papacy, or the Byzantine Empire might have used this opportunity to strike a blow against the Turks on the Aegean coast and across the Sea of Marmara in order to strengthen their control over the eastern Mediterranean. Whether the European powers were able and willing to do so is another matter. According to Pears, European "efforts made to resist [the Ottomans] were spasmodic and showed little power of coherence between the Christian states."[51] Nonetheless, despite their lack of unity and decisiveness, European states threatened the Ottomans more than did the Anatolian tribes. For instance, in 1391 Venice, worried about the growing Ottoman menace on the Adriatic Sea, gave its full support to the port of Durazzo, which was under pressure from the Ottoman army.

The last campaign of Murad is indicative of the Eurocentric direction of Ottoman geostrategy. In 1385 Murad conquered Sofia. Four years later, in the 1389 battle of Kossova, he thwarted the last attempt of the unified Balkan Slavs to arrest Ottoman expansion.[52] Although he was killed in the battle, Murad succeeded in asserting Ottoman influence over the Balkans.[53] A few years later the Europeans tried to expel the Ottomans from the Balkans through a combined sea and land attack. Hungary and France led a European coalition against the Ottomans in the hope of halting their advance by land. In 1393 Hungary, by now directly neighboring Ottoman territories, defeated the sultan's army in a battle. The Hungarian victory spurred a European expedition led by a group of elite French knights that arrived on the Danube, the northernmost Ottoman frontier, in 1396. The European army, which engaged the Ottomans near Nicopolis, was annihilated because of a tactical blunder.[54] The disastrous defeat at Nicopolis reinforced the image of the Ottomans as invincible soldiers.[55] And it paralyzed the Europeans, who for a while lost their aggressive fervor against the Ottoman Empire.

On the maritime front, European powers were unable to translate their naval superiority into a tool for containing Ottoman land expansion. For instance, in a rare display of unity Venice and Genoa defeated an Ottoman fleet near Gallipoli in 1391, but with limited effects.[56] Ottoman influence expanded by land, and European naval victories could not contain it. In fact, the Ottomans threatened Europe's maritime supremacy in the Mediterranean by controlling the shores. For example, even without a sizeable navy Bayezid I, Murad's successor, succeeded in closing the Dardanelles to navigation by building the Anatolia Hisar fort on the Asian shore. Without manning a fleet, the Ottomans controlled the key sea lane in the region. As a result, by the early fifteenth century the Byzantine capital of Constantinople was surrounded by Ottoman territories and had limited or no access to the rest of the world by sea.

By the end of the fourteenth century, less than a hundred years after their first European forays, the Ottomans had an empire that stretched across the Dardanelles and Bosphorus straits and occupied Anatolia and a large portion of southeastern Europe along the Danube. Murad alone, over the course of twenty-seven years (1362–89), extended Ottoman influence as far as Sofia and the Adriatic Sea, surrounding the once awe-inspiring Byzantine capital of Constantinople.[57] More importantly, the Ottoman Empire already had a clear geostrategy that lasted until the end of its expansion in the seventeenth century. Murad, together with his predecessors, imparted a European direction to the Ottoman geostrategy while at the same time keeping a vigilant eye on the southern frontier. Southward expansion was conditional on the strength gathered in the wealthy European regions. At the same time, the extension of Ottoman control in Europe depended on the tranquility of the southern borders. Like Byzantium, the Ottoman Empire could not afford to fight a two-front war with Europe in the north and tribal forces in the southeast.

THREATS TO THE EMPIRE

At the end of the fourteenth century the greatest threat to the young Ottoman Empire came from an unexpected foray of Tamerlane from Asia. Ottoman victories, especially at Nicopolis, "not only confirmed Ottoman control of the Balkans but greatly raised Ottoman prestige in the Islamic world."[58] In 1398, under Bayezid I's command, the Ottomans easily expanded in Anatolia, reaching the banks of the Euphrates. But the rapid Ottoman expansion in Asia was dramatically checked by the irruption on the Eurasian stage of Tamerlane and his hordes. In a rapid and violent attack Tamerlane broke through the Ottomans' southern and eastern borders, penetrating through Persia and reaching Tbilisi in 1386. In 1395 Tamerlane continued unhindered his advance westward, expanding in Anatolia, then Syria, taking Damascus and, later, Baghdad.

Devoting all his energies to defeat the Hungarian and French knights in the wealthy Balkans and the strategic Danube valley, Bayezid I initially disregarded the situation in the Caucasus and Persia. But after Tamerlane occupied the Ottoman city of Siwas, reportedly killing 120,000 inhabitants, including one of Bayezid's sons, he posed a direct threat to the center of the Ottoman Empire and could no longer be ignored. The Ottoman sultan hurriedly organized an expedition to defend the southeastern frontier against Tamerlane. On 28 July 1402 Tamerlane's hordes met Bayezid I's armies near Angora. The Ottoman troops, fighting together with a sizeable Serbian detachment, were completely defeated by Tamerlane. Bayezid survived the battle but became a prisoner of the Mongol leader for the rest of his life. After the Angora battle Tamerlane continued his advance, conquering Smyrna, held by the Knights of Rhodes.

In 1403, as suddenly as he had arrived in Anatolia and Persia, Tamerlane retreated to Central Asia to plan an invasion of China.[59] His spectacular expedition in the Middle East turned out to be a plundering raid on a grand scale, not a planned campaign to conquer new territories. Nevertheless, Tamerlane "had given the greatest check to [the Ottoman power] which it had yet received. The empire of the Ottomans which [Bayezid I] had largely increased, especially by the addition to it of the southern portion of Asia Minor, was for a time shattered."[60]

Wittek argues that the Ottoman Empire until then was only "the premature dream of an audacious ambition."[61] The Asian cities of Bursa and Nicea fell under the pressure of Tamerlane, and the Ottoman Empire "appeared to be falling to pieces in every part east of the Aegean."[62] Such an interpretation of the effects of Tamerlane's raid exaggerates the destruction of the Ottoman Empire. The sultans' empire was more than a "premature dream" because its European portion, the main geostrategic objective of the Ottomans since the time of Osman, survived Tamerlane's hordes unscathed. While the Asian borders of the empire were difficult to garrison and defend, particularly in the outer periphery of Syria and Persia, the Bosphorus and the Dardanelles were the last line of defense against an attack from the south.[63] Suggestive of the strength of the European territories was the fate of the survivors of the battle of Angora. Suleiman, the eldest of Bayezid I's sons, escaped from the battle and crossed the straits into Europe, where he and his entourage were safe from Tamerlane's attacks. The Ottoman Empire survived Tamerlane's expedition by retreating into Europe. The Eurocentric geostrategy proved to be successful.

RECONSTRUCTION

For more than a decade after the battle of Angora the Ottoman Empire was torn by civil strife. Bayezid I's sons competed for control of the remains of their father's empire. Finally, in 1413 Muhammed I defeated his five brothers and proclaimed himself the grand sultan. He inherited an empire that had lost most of its territories in Asia Minor and in the Balkans. In Asia Minor local warlords rebelled against Ottoman authority, grasping the opportunity created by the havoc wreaked by Tamerlane.

In the European territories under Ottoman control the Balkan Slavs took advantage of the Ottoman succession war and wrested some territories from the sultan's control.[64] European powers supported the local Balkan leaders by conducting small naval operations. For example, in 1416 Venetian vessels raided a few Ottoman ships in the Aegean, close to the Sea of Marmara. The ensuing battle of Gallipoli, on 29 May 1416, resulted in a complete naval defeat for the Ottomans. But as in the previous century, Ottoman defeat on the sea did not translate into a loss of territory. The Venetians were unable to extend their military success on land, while other European powers failed to organize an expedition against Ottoman territories. As a result, Muhammed could ignore the unfavorable situation on the sea and prop his weak empire by building a series of forts on the Danube. The creation of a stable land frontier in the north, thwarting European attacks, allowed Muhammed to consolidate power and rebuild the imperial organization.

Muhammed I's successor, Murad II (1421–51), brought the unruly Anatolian warlords back under Ottoman control. After he had stabilized the Asian frontier, Murad II pursued again a three-prong expansion in Europe. First, he extended Ottoman control over Greece, including the strategic Isthmus of Corinth. Second, he reconquered Serbian lands, including the city of Skopje. Finally, he pushed north, across the Danube, into the Hungarian plains.

The Danube River was of vital importance to the Ottoman Empire. It was the only natural boundary separating Ottoman lands from the European potentates in the north. According to Inalcik, "Since 1427, when the Hungarians seized Belgrade from the Serbians, the most important questions for the Ottomans was how to ensure control of the Danube. This was essential for protection of their position [in the Balkans]."[65] Belgrade, in the words of a contemporary, was "the port and the door to Hungary."[66] Located at the intersection of the Danube and Sawa rivers, it controlled access to Central Europe as well as to the Balkans.

In 1440 Ottoman forces besieged Belgrade, provoking a unified European military response.[67] The reason for Europe's unity was not Christian ideals but the strategic value of Belgrade and the national interest of Poland. Poland and Hungary, under the command of one king elected by noblemen of both countries, led the charge against the Ottomans. The Poles especially were eager to fight the Turks in Hungary rather than on Polish soil.[68] The objective was to push the Ottomans completely out of Europe, and the strategy was to assault their empire from two sides. In the south, an uprising by Ibrahim Beg of Karamen in Asia Minor, probably organized with the *beneplacet* of the anti-Ottoman coalition, forced Murad II to remove troops from the Hungarian border.[69] European forces took advantage of the temporary military weakness of the Ottomans, defeating them on the Morava River, near Nisz, in 1443. After several smaller military engagements the Europeans conquered Sofia in December 1443.[70] Worsening weather forced the Christian army to retreat to Hungary, where they wintered in relative safety close to their supply lines.

The successful 1443 campaign showed that the Ottoman army was not invincible and, above all, that it could not fight a two-front war.[71] Because of the continued instability on his Asian frontiers, the Ottoman sultan had to sign a truce with Hungary that obliged the two parties not to cross the Danube for ten years.[72] Moreover, the Ottomans agreed to return twenty-four fortresses on the northern bank of the Danube that the European forces had been unable to conquer.[73] Emboldened by the 1443 successes, however, Christian forces broke the truce. Hoping that the sultan would be unable to redeploy to Europe his crack troops involved in Asia Minor, the European army crossed the Danube and moved south. But in a show of great logistical capability Murad quickly repositioned his troops to the northern frontier.[74] In 1444 the two armies met near Varna, where the European army was defeated. Leading a small group of heavy cavalry, the impetuous Polish king Ladislaus charged straight into the Janissary line and was killed, together with most of his entourage. The resulting disintegration of the European army turned into a massacre in which the Hungarian noble elite was effectively wiped out.[75]

The battle of Varna was one of the worst European defeats in the history of European-Ottoman relations. It ended European attempts to evict the Ottomans from the continent. The physical destruction of the European army deprived European states of vital manpower for a generation.[76] But most importantly, the massacre on the Varna battlefield caused such consternation in Europe that for centuries Hungarian, German, and Polish kings had no desire to fight against the Turks.[77] "The effect of this defeat upon Hungary and Western Europe was appalling. The Ottoman Turks had nothing to fear for many years from the enemy north of the Danube."[78]

Murad II left a geographically odd empire. Ottoman territories reached from southern Anatolia to the Danube and Greece but were severed right in the middle by the Bosphorus and Dardanelles straits and by the city of Constantinople. Surrounded by Ottoman lands, and in spite of Ottoman attempts to conquer it since the thirteenth century, Byzantine emperors maintained control over this important port.

Murad's successor, Muhammed II the Conqueror, allegedly had wanted to conquer Constantinople since his childhood.[79] He began his reign by concentrating on military and diplomatic preparations for the conquest of Constantinople. The lengthy land borders of the Ottoman Empire diverted its resources, and an attack against Constantinople would have been impossible had the Ottoman army been occupied in fending off attacks in Europe or Asia. Remembering the problems caused by Ibrahim Beg's rebellions in the south and the European attack that followed from the north, Muhammed II first forced the Anatolian emirs to sue for peace.[80] Then, in order to free his forces from garrisoning the northern frontier, he signed peace treaties with Hungary, Wallachia, and Bosnia. Finally, to minimize a maritime threat, he concluded a treaty with Venice.[81] Through his diplomatic maneuvering, Muhammed II secured Ottoman borders and freed its forces.

Surrounded by land and cut off from sea lanes by the Ottomans, the Byzantine capital was left alone. In early 1453 the Ottomans laid siege to the city, subjecting it to several heavy assaults. Vastly outnumbered and outgunned, the Christian defenders were defeated, and the survivors massacred. On 29 May 1453 Muhammed II entered the ancient capital of the Byzantine Empire and transformed the Church of St. Sophia into a mosque.[82]

The fall of Constantinople was the end of the "Second Rome."[83] Yet, despite vociferous laments, no European power organized a crusade to conquer this city again.[84] On the contrary, the conquest of Constantinople bestowed respect upon the Ottomans. As Edwin Pears observed, after the fall of this city "when the world thought of Turks they connected them with New Rome on the Bosphorus. The Ottoman Turks had advanced to be a European nation."[85] The result was that the Ottoman Empire was no longer the object of crusades. It was no longer realistic to think of eradicating the Ottoman power from Europe. Coexistence became a viable policy toward the Ottomans.[86]

Maritime Power: Toward the Mediterranean and North Africa

The conquest of Constantinople marked the beginning of the second direction in Ottoman geostrategy, toward the Mediterranean. Constantinople was not only a political symbol of power but also a strategic port. In conquering Constantinople Muhammed II was pursuing a conscious geostrategy aimed at extending control over the Mediterranean and the Black Sea. He believed that with his fleet in Constantinople he could rule the world.[87]

With its natural harbor in the Golden Horn, Constantinople was a rare safe natural port and had been a strategic prize since its discovery.[88] Even the Delphic oracle called "blind" all those who settled on the Asian shore of the Sea of Marmara, not seeing the natural beauty and strategic advantages of the Golden Horn.[89] It was a commercial entrepôt through which naval traffic passed on its way from the Black Sea to the Mediterranean and European ports.

Moreover, Constantinople had a long naval history on which Muhammed II intended to found a sea empire.[90] The city dockyards, manned by Latin renegades, were well suited to building a fleet.[91] Yet a whole decade after the conquest of the city the Ottoman fleet was still unable to conduct a prolonged sea campaign. For example, during the 1463–79 war with Venice the galleys of the sultan did not win a single naval engagement.[92]

The Ottomans continued to focus on land expansion and had very limited naval success. Constantinople was mainly used to solidify Ottoman influence over key land routes that reached from the city across the straits into Anatolia, making it an important commercial and military center of control over Anatolia and the Middle East. Commercially, Constantinople was the terminus of caravan routes starting in China and India and passing through Persia, the Caucasus, and Central Asia. From a military perspective, it was easier to dispatch an army to southern Anatolia from Constantinople than from cities in Thrace or even western Anatolia. Constantinople had a similar strategic role with respect to Europe. From the city, land routes crossed into the Balkans, reaching the Danube valley and the main road to central Europe, as well as ports on the Adriatic Sea. The Roman Via Egnatia, for instance, extended from Constantinople to the Albanian coast and was a key artery for military and commercial traffic, linking the eastern Mediterranean with Italy and western Europe.

Muhammed II's strategy toward the Bosphorus straits is illustrative of the continental nature of Ottoman foreign policy. Instead of developing a powerful fleet that could patrol the Sea of Marmara and the adjacent waters, the sultan chose to control the straits through a series of forts on the two shores. On the Asian side, the Anatolia Hisar castle had been built by previous sultans. On the European side, before the 1453 siege of Constantinople Muhammed II quickly built the fortress of Rumeli Hisar.[93] These two forts, equipped with heavy cannons, allowed the Ottomans to close the straits at their will without having to build a fleet.[94] This strategy was implemented on a larger scale. Ottoman sea power was based on control over land bases and routes, surrounding the body of water and without developing a large fleet. As John Pryor, a student of Mediterranean warfare, has written, "The eventual consolidation of Ottoman maritime dominance in the eastern Mediterranean was achieved not by pitched naval battles but by a slow, relentless, and exhausting drive to gain possession of the bases and islands from which war galleys could control shipping along the sea lanes."[95]

The conquest of Constantinople was only a step in this drive to surround the sea by land. It continued during the 1463–79 war with Venice, when the Ottomans conquered "strategic islands and bases, such as Negroponte, [and] strengthened their position at sea immensely."[96] In all of these conquests Ottoman armies were the main forces, and the involvement of the Ottoman fleet was minimal. It is significant that during this and subsequent wars the Venetians feared the movements of the Ottoman army, which several times penetrated as far as Trieste, more than they feared the maneuvering of the sultan's fleet.[97]

Bayezid II, Muhammed II's successor, continued the same land strategy. He conquered several strategic ports: Lepanto, Modon, Koron, and Navarino. Ottoman soldiers laid waste to Croatia and reached even the region surrounding

Venice. Several Venetian ports, such as Durazzo, on the Dalmatian coast, were captured by the Ottoman army without substantial assistance from the sultan's fleet.[98]

The land routes reaching from Constantinople to the Balkans and the Adriatic proved to be more useful to the Ottomans than the sea lanes passing through the former Byzantine capital. By the end of fifteenth century the main Ottoman antagonist, Venice, despite a limited number of bases still manned a powerful fleet.[99] Deprived of its main support bases, in 1502 Venice signed a peace with the sultan that seriously curtailed the maritime hegemony of the Italian republic. Venice's suing for peace with the Ottomans seemed to place the sultan in control over the Mediterranean.[100]

In reality, the Ottomans were unable to capitalize on their victory over Venice. The 1502 peace with Venice was a result of the defeats inflicted upon the Venetians on land, not on sea. Moreover, the peace was almost forced upon the Ottomans because of threats to their empire on the northern and southern frontiers. The maritime expansion of the Ottoman Empire, therefore, was restrained.

This is not to say that Ottoman power did not instill fear in the European maritime powers. For instance, several Muslim ghazis took their crusading fervor to the sea, engaging in daring, albeit strategically insignificant, excursions.[101] Ottoman corsairs ravaged the coasts of the Mediterranean powers. For example, in 1544 Ottoman pirates under the command of the legendary Barbarossa reached Toulon, taking thousands of prisoners during the expedition. During their raid they plundered merchant galleys and established wealthy and powerful bases, such as Algiers and Tunis, on the North African coast.[102]

But the exploits of corsairs did not bring significant strategic advantages to the Ottoman Empire.[103] Even at the peak of their activity the pirates under Ottoman authority failed to inflict a devastating blow on Spain, France, and the Italian cities. The corsairs harassed but did not control the shipping along these lanes. Most importantly, the Mediterranean continued to prosper commercially, perhaps attracting even more adventurers seeking booty and fame. As Fernand Braudel has observed, there is "a close connection between trade and piracy: when the former prospered, privateering paid off correspondingly."[104]

Another reason for the limited strategic benefits obtained by the Ottomans from the Muslim pirates was the lack of official ties between them. The Ottoman sultan controlled the pirates only sporadically, mainly in the summer, when his fleet could coordinate activities with the pirates, and always with a great degree of uncertainty.[105] The corsairs were driven more by their desire for booty than by allegiance to the distant sultan.[106] Consequently, the sultan could not rely on them to develop the imperial framework necessary to control the major sea lanes connecting, for instance, Spain with Sicily or Alexandria.

To obtain strategic gains, therefore, the sultan could not subcontract all naval affairs to the corsairs. In the sixteenth century the Ottomans expanded in the Aegean and in North Africa, strengthening their direct influence in the Mediterranean. The vast land possessions in Europe and Asia gave the Ottomans an unmatched capability to outfit a navy: Bithynia and the Black Sea region supplied wood; Crimea and Bulgaria, iron; and sails and tents were made in Greece, Egypt, and parts of Anatolia.[107] Because of these resources Ottoman naval activity was at its peak in the sixteenth century, and almost every year the sultans equipped a very large navy in Constantinople.

The Ottomans' increased naval efforts were owing in part to their land expansion in the eastern Mediterranean. The Ottoman conquest of Egypt, described below, increased the importance of a sea link between Constantinople and the newly acquired territories. The islands of Crete, Cyprus, and Rhodes, strategically located on the most direct route to Egypt, suddenly became vital to the Ottomans.[108] Specifically, the Knights of St. John, located in Rhodes, presented a constant danger to Ottoman trade and to Muslim pilgrims headed for Mecca via Egypt. In 1522 the Ottomans attacked Rhodes, occupying it after an exhausting year-long siege.

The fall of Rhodes raised concern in Europe, especially among Mediterranean powers. But two other Ottoman conquests alarmed European states even further. First, in 1570 the Ottomans turned their attention to Cyprus. The Ottoman army conquered Nicosia and, after a long siege, the Venetian fort of Famagusta. Cyprus was located on the sea lanes connecting Greece and Constantinople with Syria and Egypt. Furthermore, together with Crete, the island was a launching pad for a potential Ottoman landing in Italy.[109] Control over Cyprus and Crete allowed the Ottomans to send a fleet to the western Mediterranean and the Adriatic without fear of losing their lines of communication with Constantinople.

Almost simultaneously with their expansion in the Aegean and the eastern Mediterranean the Ottomans escalated their military activities in North Africa. In a single decade they took over Djerba (1560), attacked Malta (1564), and occupied Tunis (1570). By the end of the sixteenth century the waters between Sicily and Tunis were under Ottoman control.[110] Spain in particular felt threatened because its communications with Sicily, an important supplier of grain, and Egypt, the route to Asia, fell under Ottoman control.[111]

The Ottomans' geostrategy in the Mediterranean seemed to suggest that they intended to assault Italy and Spain. Their control over the sea lanes in the eastern

Mediterranean and between Sicily and North Africa made a sea invasion of Europe tactically feasible. Moreover, commanding a fleet of impressive size, the sultan had the necessary means at his disposal. As early as 1537 Suleiman the Magnificent built an impressive fleet with the manifest goal of invading Italy. His naval expedition was coordinated with a land invasion led by the French king Francis I. Suleiman intended to land near Brindisi, in southeastern Italy, at the same time closing the Adriatic to Venetian shipping. However, the French king did not invade the Italian peninsula as promised, and the Ottomans diverted their naval expedition to several smaller Venetian islands and ports in the Aegean and on the Greek coast. In 1540 Venice was forced to sign a treaty giving up all of its lost possessions, but the Italian peninsula remained untouched.

In the 1560s, and especially after the fall of Cyprus and Tunis in 1570, a potential Ottoman sea invasion of Europe appeared even more imminent. Every spring in the 1560s rumors of an Ottoman invasion spread in Italy and Spain. European states followed the actions of the Ottoman fleet through an elaborate intelligence network that reported the recruitment of oarsmen in the Balkans and Anatolia, the delivery of construction material, and the movements of the ships in and out of Constantinople and through the straits.[112]

The fear of an Ottoman onslaught in the Mediterranean mobilized Spain and Venice. In 1571, spurred by Pope Pius V, who had tried for several years to forge an anti-Ottoman alliance, they joined the Holy League. The short-term goal was to save—or should it fall to the sultan's troops, to reconquer—Cyprus. The longer-term purpose was, at least on paper, to reestablish Christian control over the Holy Land.

The Holy League assembled a large, predominantly Spanish fleet. It sailed from Italy toward the Aegean in mid-1571. In October, almost by chance, the Ottoman and Christian fleets met near Lepanto, in a bay where the sultan's ships sought refuge for the winter. The resulting battle of Lepanto saw the Ottomans completely defeated. Of the more than two hundred Ottoman ships only a few managed to escape. According to some estimates, thirty thousand Ottoman soldiers were killed.

It was the most devastating defeat of the Ottomans. Braudel called it "the most spectacular military event in the Mediterranean during the entire sixteenth century."[113] The Ottoman navy had been weakened and from then on avoided European ships. While the sultans managed to rebuild their navy in the years immediately following the battle of Lepanto, they failed to restore their naval influence in the Mediterranean.[114] Most importantly, Lepanto inflicted a psychological blow on the Ottomans. Before 1571 the Ottomans had been seen as invincible, while

Spain and Venice, the two main Mediterranean powers, had been on a path of relative decline. Braudel argues that the victory can thus be seen as "the end of a period of profound depression, the end of a genuine inferiority complex on the part of Christendom and a no less real Turkish supremacy. The Christian victory had halted progress towards a future which promised to be very bleak indeed. . . . [It broke] the spell of Turkish supremacy."[115] After the battle of Lepanto the Ottomans no longer had the "arrogant impression that Christians were afraid to face them."[116]

Despite these achievements, the prevailing interpretation is that the victory at Lepanto was a great tactical victory but not a strategic success for the European powers. The lack of unity among the members of the Holy League, as well as among the other European powers, was a chronic problem that resurfaced after the victory at Lepanto. European states simply had divergent geostrategic objectives. For instance, Venice was interested in propping up its fragile commercial empire in the Aegean, not in defeating the Ottomans. If the sultan allowed Venetian merchants to trade with eastern Mediterranean ports, the Italian city had no compelling reason to expend energies to fight the Ottomans. Only two years after the battle of Lepanto, in clear violation of the agreement with Spain and the papacy, the Republic of St. Mark signed a separate peace with the sultan giving up its rights to Cyprus.[117]

The other members of the Holy League had similarly lost interest in continuing to fight the Ottomans.[118] Spain, the principal naval power, sought to restore primarily its authority over the North African coast, with, with Tunis and Algiers in Ottoman hands, presented a lingering threat to Spanish territories and shipping. In 1573 Spain conquered Tunis, only to lose it again a year later. Its main objective was not a protracted war with the Ottoman Empire but secure communications with North Africa and Sicily.

The battle of Lepanto failed to undermine the Ottoman Empire also because the Mediterranean Sea was not a key theater of action for the Ottomans and because its geopolitical importance was declining. The European naval victory was tactical and only achieved naval superiority in the Mediterranean theater. It did not, however, weaken the Ottomans' hold over their land possessions. To weaken the Ottoman Empire, the Europeans would have had to win a land victory equivalent to their victory at Lepanto. As one participant in the battle of Lepanto observed, "People believe that without the [Christian] armies on land it will be almost impossible to defeat the Turkish forces."[119] Throughout its history, including the sixteenth century, the Ottoman Empire focused its geostrategy on central and eastern Europe and, to a smaller degree, on the Middle East. The main trade routes and resource-rich areas were located in the Balkans and central Europe, and it was there that the Ottoman sultans concentrated their attention and energies. The conflict between the Ottoman Empire and Europe shifted from the Mediterranean Sea to the long land frontier from Austria to the Azov Sea.[120] Furthermore, no naval defeat could be as threatening to the Ottomans as the presence of another power on its land borders.[121] For instance, by the time of the Lepanto defeat the Ottoman Empire had a new, more dangerous enemy on its northern frontier: Russia. The reign of Ivan IV the Terrible (1533–84) had extended the Russian Empire close to the Ottoman Empire's northeastern borders. The defeat at Lepanto did not erode Ottoman land power and did not threaten the core of the empire.

The battle of Lepanto was strategically insignificant because the Mediterranean Sea itself was no longer of great geopolitical importance. In chapter 3 I examined the geopolitical shift from the Mediterranean to the Atlantic. Briefly, in 1497 Vasco da Gama discovered the Cape of Good Hope route, reaching Calcutta without crossing the Mediterranean and the territories under Ottoman control. Within months of this discovery the Portuguese navy controlled the Indian Ocean, redirecting the Europe-bound trade away from the Middle East and the Mediterranean. Six years after the first voyage around Cape Horn the Portuguese controlled Hormuz, closing the Persian Gulf, and harassed Arab shipping to the Red Sea, drastically decreasing the flow of trade to the Mamluk and Ottoman territories.[122]

The diminished geopolitical importance of the Mediterranean made it a secondary geostrategic target for Spain as well as for the Ottomans. Starting in the early 1500s the Mediterranean Sea was slowly abandoned in favor of the Atlantic by the European great powers and in favor of the Red Sea and the Persian Gulf by the Ottoman Empire.[123] As Braudel observes, "The Christians abandoned the fight, tiring suddenly of the Mediterranean, but the Turks did precisely the same, at the same moment; they were still interested, it is true, in the Hungarian frontier and in naval war in the Mediterranean, but they were equally committed in the Red Sea, on the Indus and the Volga."[124] The battle of Lepanto was the last clash on the Mediterranean between the Christian-European and Ottoman blocs.[125]

Toward Asia

Ottoman interest in the Middle East surged after the Portuguese discovery of a new route to Asia. It was a clear attempt to adjust the geostrategy to a change in geopolitics, caused by the Portuguese entry into Asia. Portuguese control over the

Indian Ocean threatened to sever the connection between the Ottoman Empire and the Asian markets. Ottoman geostrategy, concentrating on central Europe and to a smaller degree on the Mediterranean sea lanes, was no longer sufficient to maintain the position of power. In order to keep the routes to India open and maintain the flow of Asian goods, the Ottomans had to direct their attention and power southward, to Persia, Syria, and Egypt.[126] As an Ottoman geographer advised the sultan in 1580, it was necessary to open a channel "from the Mediterranean to Suez, and let a great fleet be prepared in the port of Suez; then with the capture of the ports of India and Sind, it will be easy to chase away the infidels and bring the precious wares of these places to our capital."[127]

Until the early sixteenth century the Ottomans had no reason to expand southward. High, rugged mountains, combined with the belligerent Shiites, made the conquest of Persia difficult, costly, and ultimately strategically insignificant. The Persian landscape was barren, and there were few economic reasons to control it.[128] Similarly, an expansion inland, south of the North African coast or east of the Syrian coast, was difficult and purposeless. There was no compelling reason to extend control over the caravan routes through Persia, Saudi Arabia, or the Sahara Desert. By controlling the access points of these land routes on the Mediterranean shore, the Ottomans effectively controlled the routes themselves.[129]

The Portuguese expansion in Asia changed Ottoman geostrategic priorities because it altered the routes linking Europe with Asia. In response to this geopolitical change, in the first decades of the sixteenth century the Ottomans reoriented their geostrategy toward the south (Egypt, the Red Sea, and the Persian Gulf). Initially the Ottomans merely assisted the Mamluks, rulers of Egypt, in building a navy. The Ottoman goal was in fact to keep trade routes open rather than to control an otherwise poor and weak region.[130] When the Mamluks failed to defeat the Portuguese navy and keep the sea lanes in the Red Sea open, the Ottomans expanded rapidly in Syria and Egypt. In 1514 the Ottoman sultan Selim the Grim quickly conquered the southern regions of Anatolia, defeating the Shiite Shah Ismail in the battle of Caldiran. He continued to expand in Syria, easily defeating the Mamluks near Aleppo in 1516. The Mamluk sultan was killed in the battle, but a faction of the ruling elite, led by a new sultan, continued the struggle and refused to surrender Egypt. Selim's troops therefore marched across the Sinai Desert and, through a skillful use of artillery, conquered Cairo in 1517.[131]

The Ottoman conquest of Egypt was relatively easy thanks to the corrupt state of the Mamluk Empire. But the larger geostrategic goal of weakening the Portuguese stranglehold on the Indian Ocean and reestablishing an unimpeded flow of Asian goods to Europe (and thus to the Ottoman Empire) was more elusive.[132] In

1538 the Ottomans assembled a navy in Suez comprising sixty ships and a few captured Venetian galleys and launched an extensive operation against the Portuguese. The Ottoman naval expedition succeeded in conquering some coastal areas in Yemen, including the port of Aden, and reached India undefeated. But once ashore, it failed to conquer the Portuguese fortress of Diu, in part because of the opposition of the local population and in part because of the military inferiority of the Ottomans.[133] The Ottomans returned to Constantinople via land, ending the first and only Ottoman expedition to India without success.[134] The Indian Ocean, and indirectly the Red Sea, continued to be under Portuguese control.

Along with their expansion to Egypt and the Red Sea, the Ottomans also pursued expansion toward Baghdad and the Persian Gulf.[135] The fortunes of this region were tied to conditions on the Red Sea. When Asian commerce could reach the Mediterranean through the Red Sea, the more expensive and less reliable land routes through Iraq decayed, and vice versa. In the early sixteenth century the land route through Persia increased in strategic importance since the Portuguese were harassing and blocking Red Sea shipping. In perennial rebellion, the Safavid tribes threatened the stability of the overland routes, and in 1534 the Ottomans attacked them, conquering Baghdad.[136] The assault was an apparent military success for the Ottomans, who marched to Baghdad almost unopposed. The light Safavid cavalry preferred to retreat in the face of the Ottoman army, leaving behind them a devastated landscape.[137] The Ottomans reached the Persian Gulf via Basra in 1546. In 1555 they signed the Peace of Amasya with the Persian Safavids, ending the war for a few years. The Safavid shah recognized Ottoman authority and agreed to end the guerrilla war he was waging against the Ottomans.

Nonetheless, the Ottomans did not have firm control over the region. They obtained more fame and territory but never defeated in an open battle the Persian Safavids, who, hiding in the mountains and deserts, remained a constant menace to Ottoman rule.[138] Moreover, the Portuguese controlled the Strait of Hormuz, closing the sea to Ottoman ships.[139] As a result, the route connecting Constantinople with Asia through Tabriz, Yerevan, Iraq, and the Persian Gulf was never under Ottoman control and remained unreliable. Although goods continued to flow through this route until the eighteenth century, the land connection between the Ottoman Empire and Asia was intermittent and limited.[140]

Portuguese hegemony over the Indian Ocean and thus Portuguese control over the Asian trade remained unchallenged. The Ottomans, exhausted from their attempts to defeat the Portuguese on the Indian Ocean and focused on their expansion in Europe, abandoned their hopes of reestablishing their connection with Asia.[141] From the mid-sixteenth century on, the sultans devoted enough resources to the Red Sea and the Persian Gulf to keep the frontier stable but concentrated on the more accessible and profitable Balkans.[142] Spices and other luxury goods continued to flow to the Mediterranean, but the competition from Lisbon and other Atlantic ports doomed the Mediterranean, and with it the Ottoman Empire, to relative decline.[143]

Back to Europe

Ottoman expansion into the Middle East was a failed and short-lived attempt to respond to the new geopolitical situation created by the Portuguese expansion in Asia. The main thrust of Ottoman geostrategy continued to be toward Europe. Even when fighting in Egypt or Iraq, Ottoman sultans considered Europe their main strategic prize because of its wealth and the strategic routes it afforded.[144] For example, in the mid-sixteenth century the revenues from Asia Minor were only three-fourths of those from the Balkans, while Egypt supplied even less.[145] Furthermore, central Europe and the Balkans in particular were at the crossroads of important strategic routes connecting Europe with Constantinople, the center of the Ottoman Empire.

Few European leaders appraised correctly the Ottoman threat to central Europe. For instance, Ferdinand, the Habsburg ruler of Austria, "kept repeating to his brother [Charles, the Spanish king] that the real Turkish threat was to Hungary and Vienna."[146] Yet, as described above, in the sixteenth century Charles V and his successor, Philip II, together with Venice, concentrated their anti-Ottoman efforts on the Mediterranean.[147]

In a sense, the territorial expanse of the Ottoman Empire was deceiving. The Ottomans controlled vast territories in North Africa, the Middle East, and Europe. Europe seemed to be only one of the many Ottoman theaters of actions, and in the sixteenth century the least active. Europeans considered the Mediterranean, with its extremely fluid political and military situation, to be more dangerous than the relatively stable central European front.

In the mid-sixteenth century the Ottomans reached an uneasy stalemate with Europe along the Danube River. In 1540, after the Hungarian leader John Zapolayi died childless, the Ottoman sultan Suleiman defeated the challenger to Hungary's throne, Ferdinand of Austria, and added Hungary to his possessions. In a series of wars, the Ottomans established a line of fortresses, including Esztergom, the spiritual capital of Hungary, that secured their border along the Danube.[148] After this large expansion led by Suleiman the Magnificent, his last one, the Ottomans and the Habsburgs barricaded themselves behind fortresses, leaving the Danube front relatively stable until the second half of the seventeenth century.[149]

The addition of two new fronts in the northern part of the Ottoman Empire contributed to the stability of the Danube border. The Ottoman expansion brought the sultans into contact with Poland and Russia. After the conquest of Constantinople, the Ottomans became a power in the Black Sea and expanded in Crimea and Ukraine, developing friendly relations with the local Tatars. This line of expansion clashed with Polish interests in the Black Sea and Ukraine, which were located on a key trade route that linked the markets of Lublin and Lwów with the Asian and Persian markets. The Crimean Tatars, Ottoman vassals, constantly raided border territories of the Polish kingdom. Between 1497 and 1501 and throughout the sixteenth century several skirmishes heated the frontier in Ukraine. In 1620 the frontier raids degenerated into a war between Poland and the Ottoman Empire, ending with the Polish defeat in the battle of Cecora.[150]

Until the late seventeenth century the hostility between the Poles and the Ottomans was held in check by their shared animosity toward Russia. Moscow, in fact, was recovering from its "Times of Trouble" and was extending its claims to the Cossacks in Ukraine and the Tatars in Crimea. In 1654, claiming to intervene on behalf of the Cossacks, Russia attacked Poland in Ukraine. The resulting war ended with the 1667 Treaty of Andrusovo, which gave Kiev and Ukraine to Moscow, putting Russia in closer contact with the Ottoman Empire.

The situation on the Ottomans' northeastern front heated up again in the 1670s. First, the Ottomans fought a war with Poland (1672–76) for control over southern Ukraine.[151] Once that war was over, the Ottomans engaged Russia in an exhausting conflict (1676–81) for control over the Cossacks and the Tatars. The results of both wars, however, were strategically insignificant in part because of the Ottoman focus on central Europe, rather than Ukraine and Crimea, which were peripheral to the Ottoman geostrategy. The Russo-Turkish war ended in the status quo ante, because the Ottomans, more focused on preparations for a campaign against the Habsburg Empire, fought with very limited resources.[152] Neither Poland nor Russia constituted a threat to vital territories of the Ottoman Empire. Control over Kiev and Lwów would have granted little geostrategic advantage to the Ottomans, causing only a constant conflict with Poland and a growing Russia, as well as with the independently minded Cossack and Tatar populations.

The main thrust of Ottoman geostrategy was against Austria and in particular the Danube valley. The geopolitical situation of the Ottoman Empire made this direction both necessary and desirable. It was necessary because the most threatened border of the Ottoman Empire was along the Danube. The Habsburg state "in the Danubian basin was like a spear pointed directly at the heart of the Empire."[153] The Austrian territories bordered on the wealthiest region under Ottoman control and were within easy reach of the access routes to Constantinople. The land border with Austria therefore required the Ottomans' undivided attention. All other geostrategic concerns were secondary.

At the same time, the Danube valley and especially Austria and Vienna, were crucial throughways in Europe. Since Roman times, the so-called Vienna Gate had controlled the access to western Europe from the East.[154] Once entrenched in Vienna, the Ottomans could have projected their power north (through the "Moravian Gate") to Poland, south (through the Alpine passes) to the Italian peninsula, and west (through several passes) to Prussia and France.[155] For the Ottomans, the road to Europe led through Austria. In the seventeenth century Kara Mustafa, the Ottoman vezir, argued for the occupation of Hungary and then Vienna because of their geopolitical importance.[156] Thus, Ottoman control of Belgrade and Buda was not only part of a defensive strategy of strengthening their Danube border but also a conscious effort to use them as launching pads for raids against Austria.

In the 1680s the Ottomans began to plan an attack against Austria. Ottoman military preparations were greeted by intensive diplomatic maneuvering in Europe. France, a traditional ally of the sultan, expressed interest in an Ottoman offensive against the Habsburgs in Austria.[157] To prevent the creation of a united front against the Ottomans, France pressured Poland against a pro-Austrian alliance. French diplomacy, however, was unsuccessful, and in March 1683 Poland signed an alliance with Austria stating that the security of the two powers was indivisible. The treaty also forbade the two powers to sign separate treaties with Constantinople.[158] As in the fifteenth century, for Poland it was more convenient to fight the Ottomans on its neighbor's territory when it was still assisted by other great powers than to wait until Ottoman armies were on the Polish borders. In fact, thanks to the strategic importance of Vienna, Austria had considerable facility in creating an anti-Ottoman alliance. Lwów, Kiev, and other Polish cities were not as strategically valuable to Europe, so an attack against them probably would not have produced an anti-Ottoman alliance of European powers.[159]

In the summer of 1683 the Ottoman army, led by the ambitious Kara Mustafa, reached Vienna and began a siege. From 17 July until 12 September the city defenders, outnumbered one to ten, resisted six major Ottoman assaults. On 11 September the combined Polish-Austrian army arrived near Vienna and after a bloody battle defeated the Ottoman troops. The Ottomans rapidly retreated to Hungary, while the European armies celebrated their victory in Vienna.

The land battle of Vienna succeeded where the naval battle of Lepanto failed. In contrast to the battle of Lepanto, the battle of Vienna defeated the main thrust of Ottoman expansion. Ottoman geostrategy had concentrated on central Europe since the thirteenth century, and the growth of the Ottoman Empire was fueled by its expansion in Europe. The 1683 battle put an end to this line of expansion.

The defeat in Vienna showed the geographic limits of Ottoman expansion. The distance from the main Ottoman bases in Constantinople and in the Balkans presented insurmountable logistical problems that limited Ottoman reach. Despite the good transportation framework in the Balkans and Austria, the Ottoman army reached Vienna at the beginning of autumn, leaving little time to conduct extensive military operations before winter. The Ottomans began the siege of Vienna at the end of the summer, and even without the intervention of European armies, they would have had to lift it with the onset of winter to avoid being cut off from their bases in the Balkans. "In Europe, Ottoman expansion had reached its natural limits."[160]

The logistical problem created by the distance of the spearhead of Ottoman expansion from its imperial center could have been solved, at least partially, by establishing bases in the periphery. The Ottomans, however, never colonized the Balkans or any other territory under their rule. The Ottoman army and its elite Janissary corps were based around Constantinople and did not draw the bulk of their manpower from the conquered territories. As a result, Ottoman frontiers had to be defended and expanded by an army that was recruited, trained, and supplied from the very center of the empire. While Europe was moving toward year-round national armies and a defensive system of frontier bases, the Ottomans continued to manage and expand their empire on a seasonal basis.

The battle of Vienna, therefore, was the first manifestation of Ottoman geographic limitations as much as a show of European relative power. It turned the tide against the Ottomans and strengthened Habsburg Austria.[161] Unlike the battle of Lepanto, the 1683 victory in Vienna did not destroy completely the unity of the anti-Ottoman coalition. In fact, in 1684 Austria, Poland, Venice, and Russia signed the Holy Alliance, coordinating their actions against the Ottoman Empire.[162] Venice attacked the Ottomans in the Adriatic on the Dalmatian coast, Austria on the Danube River in Hungary, Poland in Ukraine, and Russia in

Crimea. The constant pressure on their northern frontiers forced the Ottomans to fight almost every year in the 1680s and 1690s, exhausting the Ottoman military. As the war progressed, the Ottoman army left Constantinople increasingly late, limiting its ability to conduct strategically meaningful operations.[163] By the end of the seventeenth century the Ottomans had lost most of their territories in Hungary, Transylvania, and Slovenia to Austria.

In 1699 the Ottoman sultan signed the Treaty of Karlowitz. It was "one of the most noteworthy and important treaties that had ever been signed between the Ottoman Empire and the Christian states of Europe."[164] The Ottomans gave up parts of Hungary, Transylvania, and Slovenia, starting their slow retreat from Europe, which would last for another two centuries. The Treaty of Karlowitz also shifted Europe's attention from containing Ottoman expansion to managing Ottoman decline, signaling the beginning of the "Eastern Question." At the same time, by cutting the Ottomans off from important sources of prosperity, the treaty forced the sultans to change their main strategic concern from expanding in Europe to defending the empire against Europe.[165]

The decline of the empire started with the loss of European territories. Wittek writes that the slow decomposition of the empire "really concerned only its European possessions. But it has been quite rightly remarked that the losses endured in Europe, small as they might seem on the map compared to the circumference of the total, yet left the deepest impression on the whole, and the wounds which the empire received in Europe were closely followed by its decline and ruin."[166]

Geographically the Ottoman Empire was an Asian and North African empire. As Leopold von Ranke described it, "The main lines of march pursued by the Ottoman victories being three, one by sea in the Mediterranean, and two by land, in the east and in the north-west."[167] Geostrategically, however, it was a European great power. In fact, of the three "lines of march" mentioned by Ranke only the one directed toward Europe was fundamental to Ottoman power. In Europe the Ottomans both controlled their most threatened land border and expanded toward the region richest in resources. Ottomans' inability to control Asian sources of wealth, which were controlled by the Portuguese by the early sixteenth century, made their European possessions even more important. Their empire began to decline when they were deprived of access to their European territories. Cut off from its main line of expansion, the Ottoman Empire weakened; the machine of conquest relaxed, its soldiers rebelled, and the Janissaries concentrated on obtaining power in Constantinople. It was in Europe that the Ottomans built their power, and it was in Europe that they lost it.

Conclusion

Ottoman geostrategy was successful when it maintained the delicate equilibrium between pursuing control over resources and routes and pursuing security of land borders. From the thirteenth to the mid-sixteenth century the Ottomans expanded toward Europe because the most profitable and strategically important regions were the Balkans and, further up the Danube River, central Europe. It was from this region that Constantinople received the bulk of its revenues and it was in this region that strategic routes were located. At the same time, the sultans vigilantly protected their southern land borders without expanding them significantly.

When, however, the underlying geopolitical situation changed in the sixteenth century the Ottomans had a difficult time recalibrating their geostrategy. The fifteenth-century discovery of the Cape of Good Hope route shifted the axis between Europe and Asia from the Mediterranean and the Middle East to the Atlantic, diminishing the importance of trade routes passing through Ottoman-controlled areas. Ottoman sultans tried unsuccessfully to reestablish the link with Asian markets by expanding along the routes in Persia, Egypt, and the Red Sea, seeking to reach India. Their attempts to expand to the Red Sea, the Persian Gulf, and even the Indian Ocean reflected correctly the geopolitical situation. Ultimately, they were defeated by the Portuguese as well as by the inherent difficulties of expanding so far from the center of a land empire. The distances became prohibitive, and most importantly, the land borders, especially in the north with Europe, required constant attention and a large expenditure of resources.

The Ottomans devoted considerable energies to expanding their influence over the Mediterranean Sea, but their expansion there was transient and peripheral. The sultans were happy to give ample authority to local pirates and warlords who sought private gains through plunder, but with small strategic consequences. In the end, continental concerns on the northern and southern frontiers of the empire, as well as the geopolitical shift from the Mediterranean to the Atlantic, made Ottoman maritime efforts difficult and useless.

Finally, the history of the Ottoman Empire shows the inherent geostrategic difficulties of land powers. Because of their location, they are hindered by extensive land borders. Land borders, as opposed to coastal frontiers, are more vulnerable to external penetration and demand large resources to protect, garrison, manage, and expand them. The geostrategy of a land power therefore must focus on territories adjacent to its own; its strategic reach is limited by the political conditions on its land frontiers. As Nicholas Spykman observed, "A land power thinks in terms of continuous surfaces surrounding a central point of control, while a sea power thinks in terms of points and connecting lines dominating an immense territory."[168]

The rigidity of a land power's geostrategy does not necessarily mean that it is inherently weaker than the more agile sea powers. As the Ottoman Empire proved, it is far from easy to defeat or even weaken a land power. The sheer expanse of a land power makes it difficult for its enemies to evaluate its weak points. For example, for centuries Europe concentrated on the Mediterranean Sea, making few, if any, strategic gains against the Ottomans. A land power has to be defeated on its own ground, on land. The Ottoman Empire remained unchecked until Habsburg Austria (and to a smaller degree Poland and Russia), a comparable land power, exercised consistent pressure on its frontiers.

CHAPTER SIX

The Geostrategy of Ming China
(1364–1644)

On 1 September 1449, returning from an unsuccessful campaign in Mongolia, the Ming army found itself surrounded by Mongol forces near the camp of Tu-Mu. Exhausted, deprived of water and supplies, and led by incompetent officers, the Chinese soldiers panicked. Trusting in Mongol mercy, many shed their armor and ran toward the enemy lines, only to be slaughtered en masse. The Ming general, the eunuch Wang Chen, was probably killed by his own soldiers, angry at his tactical ineptitude. In the center of the battlefield a core of shock troops had formed a defensive perimeter around the Ming emperor. When the Mongol cavalry broke through the line of imperial bodyguards, they found the emperor sitting down next to his horse and waiting calmly. The battle of Tu-Mu was over. At the end of the day the Chinese army had been decimated, with casualties as high as half a million, the Ming emperor was in Mongol captivity, and Mongol troops were advancing toward Peking, China's capital. The Ming Empire was in grave danger.

Amazingly, the Ming defeat at Tu-Mu did not result in the collapse of the empire. The Mongols failed to take Peking and, torn by internal dissension, retreated into the northern steppes. But the battle of Tu-Mu marked the end of the Ming's offensive warfare and the beginning of the long decline of the Middle Kingdom. It was the tipping point of Ming China's geostrategy. After the 1449 battle Ming troops never again marched into the steppes seeking to destroy the Mongol threat and barricaded themselves behind an imposing but ineffective wall. But the northern border remained a border of tension and occupied most of the empire's attention and energies. Another result was China's retreat from the Asian maritime trade routes in the south and the opening of Asia to foreign powers.

Ming geostrategy was successful when it achieved a balance between stabilizing China's land borders and extending influence over the main maritime trade routes in Asia. This occurred during a period extending roughly from the rise of the Ming dynasty in 1364 until the 1450s, with the battle of Tu-Mu serving as a pivotal event. When in the mid-fifteenth century Ming geostrategy failed to secure the northern borders, the empire was forced to retreat from the sea, leaving a vacuum that was filled by pirates and European powers. As a result of such geostrategic choices, China began to decline in the fifteenth and sixteenth century, in particular relative to the European states, which were rapidly expanding along the sea lanes in Asia.

Why the Ming Case?

The study of Ming China is important for three reasons. First, it reiterates the absence of geopolitical determinism and the impact of policy decisions on the fate of a state. Second, the Ming dynasty left an institutional and cultural imprint on Chinese foreign policy that lasted well into the twentieth century and, some argue, continues to be influential. Third, the history of Ming geostrategy and the debates surrounding it are relevant to current discussions on China's foreign policy.

The history of Ming China reinforces my argument concerning the relation between geopolitics and geostrategy. Neither geography nor geopolitics determines geostrategy, but in order to be successful, that is, to increase or maintain a state's power, geostrategy must heed geopolitics. The geopolitical situation is a cold hard fact, not amenable to easy changes. It is simply there, defining a set of strategic options available to the state. And if a state wants to be powerful and even avoid a relative decline, it must pursue a geostrategy that reflects the underlying geopolitical situation regardless of its ideological, cultural, or other motivations.

The geostrategy of late Ming China did not reflect completely the underlying geopolitical situation and resulted in the decline of the dynasty in the short term and of China in the long term. Ming geostrategy, especially after the mid-fifteenth century, concentrated on China's northern border and abandoned control over vital trade routes in the south. It was an incorrect or at best incomplete reading of geopolitics because it failed to direct China's power to control Southeast Asian sea lanes and to protect China's resource-rich Yangtze basin. For several reasons Ming rulers altered a longstanding foreign policy that was aimed at controlling and expanding along trade routes, while protecting resources and minimizing threats on its land borders. Their fate was determined by their statesmanship, not by geopolitical factors. The study of Ming geostrategy, therefore, strongly refutes the charge of geopolitical determinism that has plagued the study of geography and politics since Mackinder.

It is important to examine the Ming dynasty because of its influence on Chinese history. As stated above, Ming rulers left a lasting imprint on China's foreign policy that the succeeding dynasty maintained well into the twentieth century and that some analysts argue continues to shape current Chinese strategy.[1] Some argue that an important inheritance from the Ming period was the revival of Confucian tradition, which translated into the dominance of civil bureaucrats over the military and above all in the renunciation of foreign trade and expansion. Partly motivated by such tradition, in the second half of their dynasty Ming rulers pursued a geostrategy of continental closure, insulating China from the world. The Manchu dynasty and Communist China continued a similar continental geostrategy, and some argue that Beijing continues—and will continue—to pursue a Confucian Ming-like foreign policy that disregards the pursuit of maritime activities. This interpretation of Ming geostrategy is flawed because it takes into consideration only one period of the dynasty. As will become evident, there were two distinct periods of Ming foreign policy, and it is at best uncertain which has had and which will have a stronger impact on the future of China's strategic direction.

Thus, the foreign policy of the Ming period can shed light on current debates on China's geostrategy. Specifically, the continued relevance of the Ming case is twofold. First, it shows that Ming geostrategy was not a product of deeply embedded cultural values peculiar and eternal to China. Rather, it was a historically specific policy that proved to be a costly mistake. In fact, the strategy of retrenchment started only in the mid-fifteenth century, while the early period of the Ming dynasty was characterized by an expansionistic policy. Contrary to widespread opinion, Ming China was not a peaceful potentate satisfied with signs of reverence and not pursuing power and influence in the region.[2] Because the late Ming foreign policy was a result of contingent decisions and not of deeply entrenched values, Chinese geostrategy has not always focused on its land borders to the exclusion of control over resources and trade routes. Indeed, if there is a lesson to be learned from China's history, it is that a "Great Wall" strategy, inhibiting China's expansion, does not work and that concentrating all efforts on a land border will not lead to gains in relative power.

Second, there are some striking parallels between Ming and present China. As did Ming rulers, the current Chinese leaders face choices between garrisoning land borders and expanding toward the East Asian sea lanes. Similar to modern China, Ming China had the potential to become a great power in East Asia and in the world because it had the necessary military capabilities, resources, and technology. Yet, Ming rulers chose not to expand toward the strategically vital trade routes in East Asia and, as a result, lost control of the sea and trade to the Europeans. For a variety of reasons examined in chapter 7, modern China is unlikely to repeat this mistake. But the study of Ming geostrategy and especially of its failures illustrates the choices and consequences faced by China today.

The Rise of China

The history of China from the Middle Ages to the mid-fifteenth century is a history of growing power. If we transport ourselves to the fourteenth or early fifteenth century, East Asia appeared on the verge of falling into a Chinese sphere of influence. China was the strongest state in the region and arguably more powerful than its European counterparts.[3] Comparisons of Europe's and China's strength and degree of development are obviously conjectural because these two political centers in general developed in relative separation, isolated by Central Asian mountains and deserts and by hard-to-navigate oceans.

The difficulty of gauging the relative growth of China is made more arduous by the traditional way of studying Chinese history. In fact, China's upward trajectory is obscured by the tendency to divide its history into dynastic cycles that emphasize its discontinuities. Each cycle is a self-contained history of rise and decline: dynasties rise to power, reach their apogee, decline afflicted by a corrupt ruling elite, and finally collapse under the pressure of domestic opposition or of foreign tribes. It appears that with each new dynasty China obtained new vitality pushing it again upward from the chaos created by the weakening of the previous rulers. In the process, however, the trajectory of China is lost, concealed by the circular rise and decline of dynasties.[4]

Such difficulties, however, should not prevent us from noticing the immense growth of Chinese power. Despite frequent dynastic upheavals, from the eighth to the mid-fifteenth century China was a rising power with an economy and a sphere of influence much larger than those of any European powers, including the Venetian, Ottoman, and Holy Roman empires. Its economy and culture flourishing, China was a source of technological marvels, including gunpowder, printing, seaworthy ships, and the compass.[5] China also played a crucial role in the international economy, importing raw materials and exporting manufactured goods. Commercial relations burgeoned on land in Central Asia and on sea in Southeast Asia. Trade between Europe and China through the Central Asian steppes began during the Han dynasty (AD 28–220) and continued with varied intensity during the Middle Ages and under the Song rulers (968–1279). During the tenth through thirteenth centuries, as a result of the growing fragmentation of Central Asia, land routes diminished in importance, and maritime regional commerce developed rapidly in Southeast Asia, reorienting China's geostrategy toward the south. As a result, China's political and economic center moved to the eastern and southern coasts, while the northwestern territories began to decline in importance.[6]

The upward trajectory of China continued during Mongol rule (1271–1368). The empire of the khans united China and Central Asia, restoring the land connection between China and Europe that had been interrupted by the decay of the Roman and Byzantine empires and the political fragmentation of Central Asia. As a result of the Pax Mongolica in Eurasia, in the late thirteenth and early fourteenth centuries land commerce between Europe and China again flourished. A European merchant observed that "the road you travel from Tana [at the mouth of the Don] to Catha [China] is perfectly safe, whether by day or by night, according to what the merchants say who have used it."[7] Probably the most famous European merchant-explorer of China, Marco Polo, also traveled along the land routes controlled by the Mongols.[8]

Sea trade also continued to grow. The Mongols inherited and expanded the powerful Song navy and maintained the Song maritime geostrategy in Southeast Asia. In 1274 and 1281 the Mongol khans sent naval expeditions comprising thousands of vessels against Japan. In 1292 they unsuccessfully attacked Java, and over the course of their rule they organized four expeditions against Vietnam and five against Burma. This line of expansion was even more impressive given the continental origins of the Mongol dynasty and its power in the steppe regions of East and Central Asia.[9]

Control over land and sea routes bestowed considerable power upon the Mongols. From the Central Asian routes the Mongols obtained large fiscal revenues that sustained their imperial structure. When their control over Central Asian routes began to weaken during Kublai Khan's reign in the thirteenth century the Mongols started to lose their grip on China. The disruption of trade in Central Asia decreased revenues, creating fiscal pressures that made Mongol military domination over China too costly to maintain. Expensive but unsuccessful expeditions such as those against Japan in 1274 and 1281 merely accentuated the fiscal strain on the Mongol Empire.[10]

The weakening of Mongol rule over Central Asia was accompanied by a corresponding decrease in sea power. In the early fourteenth century Mongol naval expeditions ended, and maritime trade decreased drastically. As a result, not only did revenues from sea trade dwindle but the supply of basic foodstuffs from the fertile south to the northern regions of China was threatened. Most of the grain, for instance, was shipped from the southern region to the north along the coast. "Beginning in the 1340s sea transport became steadily less reliable, and by the mid-1350s grain delivery to the north had virtually ceased."[11]

The decline of trade routes both on land and on sea weakened and threatened the existence of the Mongol Empire. In the process of imperial decline China became increasingly secluded from the rest of the world, and relations between Europe and China declined to their pre-Mongol level.[12]

The Rise and Peak of the Ming (1360–1450s)

By the mid-fourteenth century, as the Mongol Empire crumbled, smaller political centers prospered throughout China. The strongest developed around Nanjing, expanding gradually in the fertile region of the Yangtze River. The wealth and strategic importance of the region empowered the local warlords and rebellious groups, among them religious sects such as the Red Turbans. The Yangtze River and the surrounding region was so important that the Mongols granted official titles to the rebels in exchange for a guarantee that agricultural products would continue to flow to the north.[13] The rebels gradually wrested power from the Mongols and created a de facto new state in the Yangtze basin in the 1350s.[14]

From their base around Nanjing the rebels, led by Chu Yuan-chang quickly subdued other warlords in southern China. In 1368 they turned against the Mongol administrative centers in the north. Under the assault by Chu's troops the Mongol strongholds, including the capital, fell without putting up much resistance. The Mongols retreated slowly into the steppes, losing control over China but remaining militarily undefeated. Despite the looming Mongol presence in the north, the Mongol capital Tatu (Great Capital) was renamed Peiping (The North is pacified).[15] And Chu Yuan-chang assumed the name Hung-Wu, establishing the Ming (Light, brightness) dynasty in 1368.

The ascent to power of a native Chinese dynasty raises important questions concerning the nature of the new state. Nationalistic Chinese historians like to claim that the Ming conquest was the result of a popular peasant uprising sparked by the oppressive rule of an alien tribe (in this case, the Mongols). Moreover, they argue that Confucian ideals and Chinese nationalist aspirations motivated the uprising that led to the Ming dynasty. The logical conclusion of such an interpretation is that the new, native Chinese Ming dynasty was fundamentally different in nature from the Mongol Empire and pursued different political aims:

instead of being an aggressive imperial power, the Chinese Ming dynasty was concerned mainly with domestic policy and the development of a strong agricultural base.[16]

The novelty of the Ming dynasty has been greatly exaggerated, however. The dynastic change that occurred in 1368 was a result of an aggressive policy pursued by one group (led by Hung-Wu) with the goal of removing another (the Mongols), and nationalistic or class-based factors do not appear to have been important motivations. That Ming commanders were eager to accept Mongol officials and troops into their ranks is an indication that the new dynasty did not have nationalistic concerns.[17] The Mongols were officially subject to strict regulations, for instance, banning them from marrying fellow Mongols and mandating the Sinicization of their names, but the Ming rarely enforced these laws. Hung-Wu's main concern was the consolidation of imperial power, not the restoration of the abstract Chinese nature of the empire. Consequently, Ming China was not a revolutionary but a conservative power, in the sense that it preserved most Mongol policies. As Charles Hucker observed, "In a very real sense, therefore, Ming government was a trust inherited from the long past. It was not a process of aggressive, forward-looking advancing toward new goals, but a process of conserving what had been bequeathed by the ancestors."[18]

The continuity of the Ming dynasty applies also to China's foreign policy. Ming rulers did not drastically alter Chinese geostrategy but added only two features: a renewed vitality in foreign policy and a strong concern about the northern land border. The decay of the Mongol dynasty had led to a contraction of its geostrategy, especially toward the sea. From the founding of their own dynasty in 1368 until the mid-fifteenth century, Ming rulers pursued an aggressive strategy toward Southeast Asia on land (against Vietnam) and on sea, extending their influence over the key sea lanes in the South China Sea.

The Ming rulers were preoccupied with the northern land border. Under Mongol rule the northern frontier of China had been an internal border of a Eurasian empire, demarcating the natural environment of the steppes inhabited by nomad tribes from that of the cultivated fields of the Chinese. Once the Mongol Empire collapsed, what had been an ecological frontier separating nomads from peasants became a strategic frontier between two political entities with a history of conflict. The Mongols had been expelled from China without being defeated militarily and remained a powerful group that threatened the newly established Ming dynasty. Ming rulers were rightly concerned about the possibility of a Mongol return to China.

The northern frontier of China became a crucial region. It posed the most pressing strategic problem for the Ming rulers because failure to defend this frontier would have resulted in the loss of the most fertile and productive areas of China along the Yangtze River and, ultimately, in the toppling of the new dynasty. Every other direction of Chinese geostrategy, especially toward the sea, was conditional on the stability of the northern land border. When Ming policy succeeded in stabilizing the land frontier, China could then focus on the sea; when it failed and created a threatening situation to the political center of China, Peking, the maritime geostrategy had to be contracted.

Given the importance of the sea as a medium of Asian commerce, the key to a successful Chinese geostrategy was to achieve a balance between its maritime and continental components. As Hung-Wu and his successor, Yuang-lo, defined the vital geostrategic priorities of their newly acquired state, China had to stabilize its northern land border and revitalize a thrust toward the maritime routes in the south. As William McNeill observed, "The new dynasty combined the military policy of the southern Sung with that of the northern Sung. That is to say, the first Ming emperors set out to maintain a vast infantry army to guard the frontier against the nomads as well as a formidable navy to police internal waterways and the high seas."[19]

The two main directions of early Ming geostrategy (1368–1450) were toward the continental frontiers (the Mongols in the north, the Jurchens and Korea in the northeast, and the Central Asian tribes in the west) and toward the sea routes in the south (Vietnam, the sea lanes in the South China Sea, the Indian Ocean, and as far west as the East African coast).

The Continental Geostrategy (1368–1449)

The focus on the land frontier in the north was a constant in Ming geostrategy. From the very first day of the Ming dynasty the northern border presented a difficult strategic problem, and despite a limited success during the first century of Ming rule, it remained highly unstable. Ultimately, the failure to formulate and implement an effective policy toward the northern tribes led to the demise of the dynasty and the decline of China.

In the decades immediately following the rise to power of the Ming dynasty (1368) the emperors Hung-Wu and Yung-lo pursued a very aggressive policy toward the nomadic steppe tribes, in particular the Mongols. This strategy was characterized by military outposts on the border and inside the Mongol territory and large expeditions that penetrated deep inside the steppe region.[20]

The Mongols abandoned Peking in 1368 and from the safety of their steppe bases presented the greatest and most immediate external threat to Ming China. The deposed Mongol emperor established a new capital in Inner Mongolia, where he attracted deserters and sympathizers of his dynasty. Although he failed to organize an efficient administration that could unite the fractious nomadic tribes, the khan's military capability remained almost intact. Moreover, the flat terrain around Peking, the old Mongol capital, and along the northern frontier gave the khan's cohorts, skilled in cavalry warfare, a considerable tactical advantage over the less mobile Ming army, composed predominantly of infantry.

During his reign the Ming emperor Hung-Wu tried several times to inflict a decisive defeat upon the Mongols. He organized several expeditions outside of China in the north, seeking a military encounter with the Mongols. To some degree he succeeded in restraining Mongol aggressiveness, but overall the Chinese army was tactically unprepared to face the Mongol attacks. The limited mobility of the Ming army made it very vulnerable to the quick Mongol cavalry, which was difficult to locate and impossible to pursue. In several battles Ming soldiers assumed a defensive position behind ramparts only to be outmaneuvered by the Mongols. Such defensive tactics were later applied at the strategic level, with similar failures.

Initially Hung-Wu achieved some military victories. In 1370 Ming troops inflicted a serious defeat on the Mongols. But two years later the Mongols massacred two separate Ming armies that ventured into Mongolia. The 1372 defeat made Ming generals more prudent and defensive. From 1372 until 1387 Hung-Wu, preoccupied also with Japanese pirates and rebellious populations in the south, pursued a more cautious policy toward the Mongols.[21] The Ming army barricaded itself behind walls, rarely straying far from them. Such a defensive posture guaranteed a limited stability because the Mongols could not evict Ming troops from their defensive frontier outposts and could only raid Chinese territories. But this defensive tactical strategy emboldened the Mongols, who over the 1370s and 1380s launched several incursions that often resembled full-fledged invasions deep inside China.[22] Immobilized by their own walls, Chinese generals lacked the tactical mind-set and the means to respond.[23]

A decade later, feeling safer on his imperial throne, Hung-Wu embarked on a more aggressive policy. In 1387–88 Ming generals conducted two extensive campaigns against the Mongols. While the first was inconclusive, the second resulted in a great victory of the Chinese army over Mongol forces led by Toghus Temur, Genghis Khan's successor.

The military success had momentous political consequences in China. The victorious Ming generals achieved enormous prestige back in China, and their fame and popular support matched, if they did not eclipse, the standing of the emperor. Such a situation presented a threat to Hung-Wu's hold on power. In a drastic move the Ming emperor, whose "desire to seize, wield, and defend power" characterized most of his rule, purged the military nobility, imprisoning, executing, or exiling the top officers.[24] The results were similar to those caused by Stalin's 1930s purges in the Soviet Union: the army was left with no commanders competent to defend the country from external threats.

Apart from weakening the military capability of the Ming Empire, Hung-Wu's purge established the conditions for future domestic strife. The emperor filled the top military echelons with members of his immediate family, among them his twenty-six sons, who took over the command of border regions and armies. Each prince administered a region and controlled a locally based army, creating powerful centrifugal tendencies. This military system gave Hung-Wu greater control over the administration and the army, but it also created the conditions for a clash among the various princes.[25] When Hung-Wu died in 1398, the local princes, free from the constraints imposed by imperial authority, unleashed a civil war that lasted four years. During this conflict among the different contenders to the throne China had no meaningful foreign policy.

The civil war ended in 1402 with the victory of the Prince of Yen over the imperial troops, based in Nanjing. The prince was the commander of China's northern region, bordering with the Mongols, and once installed as the new emperor Yung-lo (Perpetual happiness) he moved to Nanjing. Because of his military upbringing, fighting the steppe tribes, and his geographic base of power in the north, Yung-lo gave a new impetus to Ming foreign policy toward the Mongols. The new emperor and his generals were well trained in cavalry tactics, reflecting the knowledge they had acquired while defending the northern frontier from Mongol incursions.[26] Tactically, Yung-lo's troops embraced a steppe way of warfare and were more similar to the Mongols than to the main Ming army. Like the Mongols, the northern army was highly mobile and capable of living off the land, and by leveraging these advantages it had defeated the southern-based imperial army, composed mostly of infantry, which was slowed down by long supply trains. Moreover, in contrast to Hung-Wu's detached attitude, Yung-lo, first as a prince and later as emperor, personally led his troops into battle. Such close relations with his forces not only increased troop morale but also prevented the rise of ambitious generals and princes. Yung-lo's tight control over the military, in other words, avoided the establishment of centrifugal centers of power that had led to his own rise.

Yung-lo's geographic base of power in the north also invigorated Chinese foreign policy toward the Mongols. Yung-lo rose to power thanks to the support he had garnered in the northern frontier region, and he was most comfortable ruling surrounded by his acolytes. An indication of his preference for the northern region was the fact that he spent most of his time in Peking, limiting his stays in Nanjing to a few years at the beginning of his reign and during moments of unrest in or external threats to southern China. During most of Yung-lo's reign China had two official capitals, Nanjing (the original cradle of Ming power) and Peking (the old Mongol capital and Yung-lo's base), but in 1420 Yung-lo officially transferred the capital to Peking, the core of his personal power and the main base of military expeditions against the Mongols.

The move of the capital to Peking signaled an intensification of Ming policy toward the Mongols. Based in Peking, Yung-lo spent most of his time as emperor leading military expeditions in the steppes. This was a positive development for China because the greatest threat to China's security came from the north. As a Chinese saying aptly put it, "While northern barbarians are not obedient, China cannot sleep in peace."[27] Despite the increased aggressiveness, Yung-lo maintained China's traditional reluctance to expand to the arid northern steppes. His projections of power outside of China's northern border were meant to stabilize the border, not to annex more territories.[28]

China's defensive continental strategy was a prelude to an offensive maritime one. The stabilization of the continental frontiers, and the resulting territorial security of China, was a conditio sine qua non for the pursuit of control of trade routes in the south and for the protection of China's center of resources in the Yangtze basin. In fact, the failure to stabilize the land border prevented China from projecting power toward the trade routes of Southeast Asia.[29] The reinforced fortifications in the north and the expeditions in Mongolia were an attempt to redirect Chinese expansion away from the steppes and toward the more lucrative south. As Owen Lattimore argued, the Great Wall, like any imperial boundary, such as the Danube for the Roman Empire, also "represents the limit of growth of an imperial system." China simply realized that expansion in the north was militarily difficult, strategically useless, and politically dangerous. While in the south the ecological similarity with mainland China did not create a natural boundary, in the north there was a clear natural line dividing the agricultural China from the steppes. The Great Wall served to reinforce the ecological distinction that translated into political differences. "The steppes of Mongolia and western Manchuria and the high wastes of Tibet denied the primary Chinese economic requirement of intensive, irrigated agriculture." Northward expansion would have created a completely different society, based on the nomadic life of the steppe rather than on the settled life of an agricultural society. "If carried too far it began to create at the margin a different kind of society. The state itself . . . had therefore a constant—though not always dominant—reason to restrain its borderers from venturing into the steppe." The purpose of Ming geostrategy toward the north, therefore, was both to defend against the Mongols and to prevent further Chinese expansion. "An imperial boundary . . . has in fact a double function: it serves not only to keep the outsiders from getting in but to prevent the insiders from getting out."[30]

But the move of the capital to Peking from Nanjing, meant to strengthen the northern frontier, also weakened the Ming state. First, placing the administrative and political capital within the reach of Mongol forces exposed the center of the Ming state to their raids. The first Ming capital, Nanjing, was not only in the middle of a fertile region but also distant enough from the northern steppes to make it safe from Mongol incursions.

Second, the move limited Ming interest in maritime activities. Nanjing was considered "a good location for a base from which military expansion could proceed in every direction."[31] Peking instead was forty miles from the northern border, where later the Great Wall would be built, and near the Nan-k'ou Pass, which was the main access point from Mongolia to China.[32] The strategic focus of Peking was the north, not the south, and would contribute to the decline of Ming maritime geostrategy.[33]

Finally, the new capital Peking was dependent on the southern regions for its economic survival. Most of the food had to be supplied from the Yangtze region via coastal lanes and later through the Grand Canal, which connected Nanjing to Peking.[34] Since a disruption in the transportation network delivering the food could threaten the survival of the capital and the imperial court, Peking had to devote resources to the protection of these routes. Moreover, shipping food from the south to Peking absorbed an increasingly large part of the Ming navy, which especially after the opening of the Grand Canal lost its oceangoing capabilities.

Initially the perils of moving the capital to Peking were outweighed by the benefits of a more aggressive foreign policy toward the Mongols. The cautious policy of Hung-Wu was replaced by a more assertive strategy that projected China's military power well beyond its northern border in frequent expeditions.[35] In the years 1410–24 Yung-lo personally led five expeditions against the Mongols, in 1410, 1414, 1421, 1423, and 1424. In each campaign the Chinese army was mobilized in Peking in the spring, crossed the Great Wall sometime in the summer in a "search and destroy" mission against the Mongols, engaged in several but never decisive battles with the nomad tribes, and finally, claiming victory, returned to the relative safety of the Peking walls in the fall.[36]

The expeditions in the years 1410–24 were temporary thrusts of power, not a grand campaign to annihilate the Mongols and conquer their territory. It is true that they succeeded in establishing a moment of stability on the frontiers by taking the war directly to Mongol territory. In the first years of Yung-lo's reign Ming armies also built forts and military encampments outside the Chinese line of walls on the border. The presence of Ming soldiers in the Mongol steppes, combined with frequent large-scale expeditions, created a buffer zone that permitted an in-depth defense.

Nevertheless, there were problems with such a policy. The stability it generated was not sustainable in the long term. The size of the Ming armies organized for Yung-lo's expeditions imposed a serious burden on Ming finances and on the emperor's time without achieving lasting strategic gains. Despite Yung-lo's training in the highly mobile frontier warfare, the Ming army was unable to inflict a decisive defeat upon the Mongol cavalry. Moreover, toward the end of Yung-lo's reign the in-depth defense was slowly abandoned. Part of the territory of the buffer zone was given to the Urianghai Mongols in return for their support of Yung-lo during his struggle for power in China. The gradual abandonment of outposts in the steppe region put the Ming on the defensive. In fact, "experience . . . had demonstrated that it was impossible to secure northern China without powerful influence, if not control, over the steppe. That influence depended on the maintenance of bases and striking forces along the steppe margin."[37]

As a result, the original strategy, which involved a constant military presence outside China's walls, was replaced by a more defensive policy reminiscent of Hung-Wu's period. Ming troops chased Mongol cavalry detachments outside the Chinese walls, but after an engagement they retreated back into the safety of their frontier posts. Such tactical timidity was owing partly to the increasingly large supply trains that followed Chinese troops. It was a sign of a growing "sclerosis" among Ming generals. As Edward Dreyer observed,

> In the 1370's and 1380's Ming generals had led cavalry armies that could overtake and engage the nomadic Mongol hordes. Even in the civil war of 1399–1402, Yung-lo's army, because of its superior mobility, was able to escape annihilation by larger forces and ultimately to defeat them in detail. In contrast, the armies of Yung-lo's Mongolian campaign became increasingly bound to their supply trains as the reign progressed, and their complexity in organization grew correspondingly.[38]

In part, however, the failure of Yung-lo's Mongolian wars was owing to the political nature of the conflict. The Mongols were an extremely difficult target to locate and destroy. Not only were they very mobile but they lacked a unified leadership and were divided among several tribes. Ming rulers, including Yung-lo, encouraged the divisions among the various Mongol groups, pitting the no-madic tribes against one another through alliances and bribes. The motivating idea was to divide and rule: a divided enemy was weaker and less able to conduct a coherent strategy. But the weakness of the Mongols proved also to be a strength. The destruction of one horde or the defeat of one leader did not undermine irreparably the other Mongol tribes. On the contrary, it merely created the space for the rise of another tribe or leader. Ming policy therefore prevented the rise of a center of Mongolian power that could be attacked, destroyed, or even negoti-ated with.

Moreover, the Ming approach toward the Mongols never removed the funda-mental cause of Mongol raids, the lack of trade with China. The conflict with the Mongols was "political rather than territorial."[39] The steppe tribes were not inter-ested in extending control over more territory; they were only interested in obtain-ing access to Chinese markets. Trade between China and the steppe tribes had been vibrant for several centuries prior to the Ming dynasty. The Mongols had obtained grain, iron, and manufactured products in exchange for steppe horses (which, ironically, China needed to fights against them). While China could afford to stop trading because of its wealth in the fertile regions in the south, for the Mongols lack of trade with their southern neighbors spelled economic disaster. Consequently, when the Chinese court limited, often blocking, all trade with the Mongols, whom they considered to be culturally inferior enemies, the nomad tribes had to resort to plundering raids against the Chinese. China's prohibition of trade with the Mongols, spurred partially by the growing enmity between Peking and the steppe tribes and partially by an increasing aversion to com-mercial interactions with foreigners, exacerbated its relation with the northern tribes.[40]

The example of Ming-Jurchen relations suggests a similar connection between commercial isolation and war but a more effective Chinese policy. At the begin-ning of the Ming dynasty China had no established relations with the Jurchens, in Manchuria; there were neither fortifications nor system of tributes and com-merce. In contrast to his policy toward the Mongols, the emperor Yung-lo re-frained from organizing military expeditions against the Jurchens and pursued diplomatic relations with them, seeking to bring them closer to China rather than to the Mongols or the Koreans.[41] To achieve this, he allowed and encouraged commercial relations. Such a policy of free trade, however, did not mean that Ming rulers were not interested in controlling the Jurchen tribes and their land. An aggressive policy of Sinicization accompanied trade, establishing a strong cultural and political rapprochement between China and the Jurchens.

There were serious long-term drawbacks to such a policy. The economic and cultural closeness between the Chinese and the Jurchens contributed to the rapid development of the latter's administrative skills. The Jurchens were able to coalesce the various Manchurian tribes, and by the early fifteenth century Manchuria was a strong unified power. This political unity allowed it to counterbalance Peking by forging alliances with Korea and even the Oyirats, in Central Asia. Nonetheless, in the early Ming period (1368–1420s) the more amicable China-Jurchen relations allowed Peking to project power toward other regions, such as Southeast Asia and the Indian Ocean.[42]

Despite Yung-lo's attempts, the Mongols remained the principal threat to China's political existence, and the northern frontier continued to be a "border of tension."[43] The momentary stability achieved by Yung-lo's aggressive policy and frequent expeditions gradually disappeared. After his death in 1424 Peking's policy toward the northern frontier slowly changed and lost its aggressive élan. Yung-lo's successor even ordered the transfer of the capital back to Nanjing, in large measure because of a lack of interest in the northern frontier. His sudden death prevented the execution of the order but did not arrest the decline of Ming interest in an aggressive policy toward the Mongols. Instead, Peking pursued a policy of hermetic separation between the steppe nomads and China. China's military inactivity in the second half of the 1420s and in the 1430s allowed the Mongols to strengthen. It was during this period that the Ming court began reinforcing the line of fortifications in the north, turning it into what would be known as the Great Wall. According to Morris Rossabi, "A less activist foreign policy was pursued, and the renowned 'Three Yangs' who dominated government in the late 1430s and early 1440s sought stability and peace though limitations on foreigners."[44]

The élan of the Ming continental geostrategy abated during the 1420s, but not until the mid-fifteenth century, after a bloody Ming defeat, did a clear break with Yung-lo's policy occur. In 1434 Toghon, the leader of the Oirat region, united the petulant Mongol tribes under his leadership. His son Esen continued the process of political unification, at the same time hardening Mongol policy toward China. Tensions rapidly increased in the 1440s, and there were more frequent Mongol raids into China. In 1449, in response to a large-scale attack by Esen, Peking assembled a powerful army and organized an expedition to Mongolia led by the

Ming emperor himself. Following a traditional pattern, after an extensive foray outside of the Chinese line of fortifications the Ming emperor began to retreat without having encountered the bulk of the Mongol forces. In order to wait for the wagons full of luxuries that the emperor and his eunuchs had taken on their expedition, the army camped outside a fortified post near Tu-Mu. Catching the Ming army utterly unprepared, the Mongols attacked and annihilated the army, taking the emperor prisoner. The Mongols were stunned by their own victory and instead of capitalizing on it slowed their advance into Chinese territory, giving the Ming court time to reorganize and choose a new emperor.[45] When the Mongol leader Esen pressed on, putting Peking under siege and hoping to use his imperial hostage to blackmail Ming authorities, the Chinese were already prepared for the defense. Indifferent to the fate of their own emperor, the Chinese repelled the attacks of the Mongols. Unprepared for a protracted siege, Esen retreated to Mongolia. He died shortly after the siege of Peking, killed by members of his own court, and Mongol unity dissolved again.

The defeat at Tu-Mu was a pivotal moment in the history of Ming geostrategy.[46] Even though the campaign resulted in a Mongol retreat and eventually the end of Mongol unity, the Ming court lacked the resolve to pursue an offensive strategy against the steppe forces. The shock caused by the massacre at Tu-Mu, compounded by the humiliation of having an emperor taken hostage and by the siege of Peking, led to a radical reassessment of Ming policy. After Tu-Mu a more defensive mind-set prevailed over Yung-lo's aggressive policy toward the Mongols. Peking completely abandoned the military outposts and fortifications that, together with frequent expeditions, were part of a buffer zone established during the reigns of Hung-Wu and Yung-lo beyond the Great Wall.[47] As observed above, the buffer zone did not annihilate the Mongols, but it created some stability on the northern frontier. Some Ming strategists appreciated the dangers of abandoning the buffer zone. In the second half of the fifteenth century a Chinese strategist encouraged the emperor to reconstitute the buffer, where "in the past we had the distant garrisons as an outer barrier. Later they were moved back within, and now the fence protecting the capital and the east and north is thin and unsubstantial indeed."[48] The capital of the Ming Empire, Peking, was only about sixty miles from this "thin fence," the Great Wall, and the fear of a Mongol attack would continue to consume China's attention and energies.[49]

The abandonment of the buffer zone led to "the second, and defensive, phase of Ming frontier operations, which centered on the Great Wall."[50] It was a last-ditch effort to arrest the nomads' assault.[51] This strategy, based on the increasingly elaborate line of fortifications and walls, was a sign of China's weakness relative to the Mongols and its inability to formulate an effective strategy to deal with them.[52]

The Great Wall represented the wrong tactical solution to the problem posed by the Mongols. Positional warfare simply could not protect against a highly mobile enemy like the Mongols, and the Great Wall was as ineffective a defense as the Maginot line in the twentieth century. The construction of the wall continued from the fifteenth century until the fall of the Ming dynasty in 1644, but the threat posed by the Mongols, and in the late sixteenth century by the Manchus, never diminished. The costs of this constant defensive system were enormous, while the military effectiveness was minimal.

The chronic instability of China's northern land border could have been avoided. As F. W. Mote observes, the Ming court "allowed the entire relationship with Mongolia to be reduced to a static border defense policy—costly, inefficient, mismanaged, and destructive of the texture of Chinese life in the border zone. It was the greatest failure of Ming statesmanship."[53] Because the war with the nomads was more political than territorial, Peking could have dealt with the conflict by political means, namely, by allowing trade with the steppe tribes. But bureaucratic pressures against commercial relations with foreigners, especially the culturally inferior "barbarians" in the north, were powerful and shaped Ming policy toward the steppe tribes.

Thus, the continental geostrategy of the Ming was a failure. Stability was achieved for a brief period during the early Ming dynasty, but after the mid-fifteenth century the northern frontier became the preeminent strategic problem for Peking. It absorbed attention and resources, and above all, it made expansion toward the south a secondary strategic objective.

Northern Vietnam (Annam)

In the first decades of the Ming dynasty, when the northern frontier had a modicum of stability, China also focused on the south. Specifically, Ming geo-strategy concentrated on two regions that had great strategic importance, Annam (modern northern Vietnam) and the sea. Annam was on the border with China, and its political instability was a threat to China's southern regions. Moreover, control of Annam's coast meant influence over the busy sea lanes in the South China Sea, connecting China with the Malacca Strait. The Ming geostrategy toward Annam, therefore, was part of a more ambitious policy toward the sea. Peking aggressively tried to extend its influence over the sea lanes of Southeast Asia and as far west as the Indian Ocean and the East African coast. Control over these sea lanes would have given Peking unmatched power in Asia, protecting China's resource-rich center in the Yangtze region and its access to the markets of India, Persia, and, indirectly, Europe.

China's expansion in Annam was a brief and ultimately disastrous episode during the Ming dynasty. The importance of Annam was twofold. First, the conflictual relations between Annam and its neighbors (such as Champa, modern southern Vietnam) destabilized the region, with potentially dangerous spillover effects to southern China, already beset by a rebellious population. The disputes between Annam and its neighbors threatened to draw Chinese generals and armies into local conflicts, limiting their ability to impose imperial order over southern China. Second, Annam, and Southeast Asia in general, was a strategic region situated on important land and sea routes connecting China's ports with the Indian Ocean. The southern routes, passing through Vietnam, provided China with "luxury goods, which included pearls, incense, drugs, elephant tusk, rhinoceros horn, tortoiseshell, coral, parrots, kingfishers, peacocks," and other goods.[54] Chaos in Southeast Asia would have disrupted trade flows through the region. A potential vacuum in Annam, therefore, was a direct threat to China's territory and its vital trade routes.

China's strong cultural and political influence upon Annam mitigated its potential threat. Annam was the most Sinicized area of Southeast Asia. Although officially independent, it was under the authority of the Ming emperor. From 1368 until the early 1400s Ming China de facto controlled its southern neighbor, considering it a vassal of sorts. China's policy during this period reflected its relationship with Annam. In the 1370s through the 1390s Annam was entangled in a conflict with Champa, to its southeast. The possibility that the conflict might spread to southern China worried the Ming court, but Emperor Hung-Wu maintained a posture of impartiality, appealing to Annam and Champa to cease fighting. The apparent neutrality, however, ended when China needed supplies of food for its frontier garrisons engaged in pacifying rebellious areas in the south. Annam was forced to deliver grain and other foodstuffs, demonstrating its political dependency on Peking.

China's attitude of neutrality toward Annam was in part owing to its preoccupation with the Mongols in the north. In 1373 Hung-Wu enunciated a policy of nonaggression toward China's neighbors, including Annam and Champa. Southeast Asia, together with other regions, were "separated from us [China] by mountains and seas and far away in a corner." The fact that the region could not sustain Chinese troops and population, if they chose to settle there, magnified its political and economic insignificance for China. According to Hung-Wu, "Their peoples would not usefully serve us [the Chinese] if incorporated [into the empire]." But Hung-Wu's pledge of nonaggression had an important caveat that left open the possibility of Chinese intervention outside its borders: "If they [China's neighbors] were so unrealistic as to disturb our borders, it would be unfortunate for them." He was especially concerned about the northern barbarians who threatened China's frontiers, and he argued that the best forces ought to be sent to the north, not to Annam or Southeast Asia.[55]

Hung-Wu's policy was more a statement justifying Chinese inability to conquer the south than a definition of an eternal strategic principle for China. In fact, behind the appearance of political neutrality and cultural superiority, China did not enjoy absolute control over Annam. Apart from agreeing to be part of the tribute system, Annam did not consider China the center of the universe, and in fact it thought of itself as a rival, if not equal, imperial power. Similarly, although it maintained an appearance of unmatched imperial superiority, China tried unsuccessfully to establish relations with Burma in order to counterbalance Annam. The ambiguity of the relationship between China and Annam kept Chinese leaders alert, and indeed, Hung-Wu's admonition notwithstanding, China never abandoned its thrust southward toward Annam.[56] If it was limited in the first decades of the Ming dynasty, it was because of domestic problems: the first Ming emperor had to devote most of his attention and strength to consolidating power and imposing order on a society in chaos.[57] But even during such a delicate time relations with Southeast Asia often degenerated into violence because of border incidents.[58]

Chinese involvement in Annam intensified in 1400, when an usurper killed Annam's legitimate emperor. When in 1405 the usurper ambushed and massacred a small armed Chinese expedition whose goal was to restore the legitimate heir to the throne, Yung-lo ordered the invasion of Annam. After the fall of Annam's two capitals, the Chinese generals in charge of the expedition annexed it as a province of China. However, "the idea that the sinicized Vietnamese would welcome incorporation in the only civilized empire—no matter how reasonable from the perspective of Nanjing—ignored the strength of the historical traditions of Vietnamese independence and their hostility toward Chinese overlordship."[59] The Vietnamese rallied behind several prestigious generals and waged a relentless guerrilla war against the occupying Chinese troops. Yung-lo, increasingly more occupied with his Mongolian campaigns, paid scant attention to the situation in Annam. In 1426 the Vietnamese soundly defeated a Chinese army sent to recapture the town of Tra Long. The military success incited nationalist feelings that led more men to join the anti-Chinese guerrilla army. In the months after the victory at Tra Long the Vietnamese dealt several setbacks to the Ming armies. Finally, in 1427 the Chinese army was on the defensive and began to retreat from Annam. To save face, the Ming emperor recognized the usurper as the legitimate ruler of Annam and took his own troops back to China. China's thrust toward continental Southeast Asia was over.[60]

The failure in Annam preceded by two decades the disastrous Tu-Mu defeat in 1449, but it had a similar effect on Ming geostrategy. It reined in the imperial push of Peking. Ming civil servants, traditionally opposed to foreign contacts and expansionistic policies, used it as "a rhetorical weapon against future large-scale military undertakings."[61] But the Chinese occupation of Annam was also unsustainable because of the costs, which, especially in light of the growing menace on the more sensitive northern frontier, became prohibitive. Given the limited resources, Peking chose to sacrifice its control over Annam in favor of a stronger defense in the north. A failure in Annam was less dangerous than a failure on the northern border. In the end, Hung-Wu's admonition not to waste energies in the south when the northern frontier was threatened turned to be prophetic.

The Maritime Geostrategy

The failure of Ming policy toward the Mongols, and to a smaller degree toward Vietnam, curtailed China's maritime geostrategy. Continuing a long maritime tradition, the early Ming emperors pursued an ambitious sea geostrategy that extended China's influence over South Asian waters and even the Indian Ocean. However, by the mid-fifteenth century the unstable land borders in the north not only absorbed an increasingly larger portion of government revenues but also threatened the very heart of the empire, the capital, Peking. In light of such a situation, maritime expansion was prohibitively costly in both fiscal and strategic terms. China retreated from the sea, but the ensuing limitations imposed on the Ming maritime geostrategy had devastating effects on China's ability to control key trade routes, to protect its coastal regions, and, from the sixteenth century on, to respond to the European challenge in Asia.

Until the mid-fifteenth century China was a rapidly growing naval power within a busy Asian network of maritime trade. The rise of Chinese maritime power dates back to the fifth century, but it was only in the twelfth and thirteenth centuries that the Middle Kingdom established supremacy on the Asian seas. The Song dynasty (960–1279), partly because it was cut off from key land routes in the north and west, developed a powerful navy that by the early thirteenth century was the best in the Indian Ocean.[62] When the Mongols invaded China in the north, the Song rulers put up a stiff resistance in the south thanks to their naval strength. To control southern China, the Mongols had to defeat the Song on the sea. Helped by deserters and Chinese merchants eager to gain commercial advantages, the Mongols built their own fleet and raided the coastal areas, eventually forcing the surrender of the Song rulers.[63]

The Mongols inherited the Song navy and, despite their continental origins and their power base in the northern steppes, continued to expand on sea. In 1274, in an incessant quest for power, they launched a naval expedition of about one thousand ships against Japan. Although the Mongols landed in Japan and won a few battles, a violent storm (called a "kamikaze" by the Japanese) wrecked their fleet and forced them to retreat to China. In 1281 Kublai Khan organized another expedition against Japan on an even larger scale. But even though the Mongol fleet had forty-four hundred ships of various sizes, it was completely destroyed by a monsoon wind.[64] Twelve years later, however, the Mongols again rebuilt their fleet and with more than one thousand ships attempted an invasion of Java in order to safeguard the passage of trade through the Malacca Strait.[65] The invasion was unsuccessful, but the Mongol navy extended Chinese influence as far as Hormuz, in the Persian Gulf.

China had an impressive navy that was probably stronger than any European fleet. In fact, the Song and Mongol navy was an object of marvel among European visitors. They reported back to Europe in great detail about the technical specifications of Chinese ships, which were much larger and more sophisticated than the average European vessel. According to G. F. Hudson, "The largest had fifty or sixty cabins apiece, four or even six masts, double planks and watertight compartments, towed two or three large boats and carried some ten small ones, had crews of 200 or 300 and took as many passengers, grew vegetables on board and carried as much as 6,000 baskets of pepper in the cargo."[66]

The rise of Chinese naval power continued with the Ming dynasty despite some early reservations on the part of the court about pursuing commercial activities. The Ming inherited from the Mongols river, coastal, and oceangoing fleets, as well as navigational skills and shipbuilding techniques, but also a certain anxiety about commerce, and maritime trade in particular. The first Ming emperor, Hung-Wu, officially prohibited maritime trade and even foreign products such as perfumes. In 1397 an edict only allowed foreign ships to visit Chinese ports to pay official tribute to the emperor.[67] This aversion to maritime commerce was owing in part to the strengthening of Confucian beliefs. Confucian imperial advisers argued that China ought to maintain its agricultural character and avoid commercial undertakings. Maritime trade seemed to be particularly influential in detaching the Chinese population from their land and to bring them into contact with foreigners, considered to be culturally inferior.

But Confucian ideals were not the only reason, and certainly not the predominant one, for the diffidence toward sea trade.[68] After all, Mongol emperors, who did not surround themselves with Confucian advisers, also limited commercial enterprises, outright forbidding Chinese merchants from going overseas in 1309.[69] Rather than reflecting Confucian beliefs, these bans stemmed from the need to strengthen the imperial government by centralizing control over the commercial fleet. In the early Ming dynasty, during Hung-Wu's reign, China was in a similar situation that required the consolidation of its imperial power, which was weakest in the coastal areas. There, commercial activities made the populations wealthy and consequently less pliable to central authority. The early bans on trade, therefore, prohibited private, but not state-run, commercial activities. A state-run commercial system, in fact, limited the establishment of private ventures, which could have presented a threat to imperial authority. The emperor was the highest individual, in charge of heaven's mandate, and consequently all activities, whether military or economic in nature, had to originate with him.

In spite of such restrictions on private commercial enterprises, maritime trade continued to prosper and increased throughout the fourteenth century. Many Chinese, for instance, settled in Sumatra or in other regions of Southeast Asia in order to continue their commercial businesses. Moreover, the revenues obtained from taxing imports and exports in the various ports of southern China were an important source of state income under both Mongol and early Ming rule.

Most importantly, despite the bans and the preoccupation with internal politics, Hung-Wu enlarged the fleet. At the peak of his reign China's navy numbered between 3,500 and 6,500 ships.[70] The navy was used for coastal commerce as well as for trade with India, East Asia, and even Africa. But China's sea power was limited to the coastal waters and the South China Sea. Farther away, along the Asian sea lanes, Ming maritime influence was minimal. In the Indian Ocean China's involvement was limited because until the late fourteenth century Arab merchants controlled the main sealanes. Spearheaded by traders, from the twelfth through the fourteenth century Islam spread throughout the region, reaching even Malaysia and transforming the Indian Ocean into a "Moslem lake."[71]

The Ming navy, although large and powerful, could not dislodge Arab ships from the Indian Ocean. The logistical difficulty of projecting naval power so far away from home was enormous, and Hung-Wu did not organize any expeditions to establish a Chinese presence in those waters. China's limited participation in maritime commerce in the Indian Ocean, however, does not mean that the exchanged goods were not strategically important. The products traded, in fact, were not only luxury goods such as porcelain or silk but also key resources that China needed: "horses, copper ores, sulphur, timber, hides, drugs, and spices, not to mention gold, silver and rice."[72]

It was only with Hung-Wu's successor, Yung-lo, in the early fifteenth century, that Ming geostrategy turned toward the Indian Ocean, marking the apogee of Chinese sea power. In the years 1405–33 Yung-lo organized seven spectacular expeditions that took Chinese merchants as far as the East African coast and the Persian Gulf. These expeditions extended Chinese influence over the South China Sea, the Malacca Strait, and the Indian Ocean and established a small Chinese commercial presence in East Africa.

In the first decades of the fifteenth century, therefore, China was well on the way to becoming the predominant Asian, and possibly global, sea power. In 1433, however, the sea expeditions suddenly ended, and Peking prohibited the further involvement of Chinese subjects in maritime affairs, stopping not only state-sponsored but also private maritime commerce and shipbuilding.[73]

In light of China's later isolation, the naval expeditions to the Indian Ocean led by Cheng Ho, a powerful and ambitious eunuch at Yung-lo's court, seem bizarre. Chinese and Western historians have proposed a variety of explanations for them. The traditional Chinese explanation is that the purpose of these expeditions was to find an emperor who had been dethroned by Yung-lo during the civil war that followed Hung-wu's death. The deposed emperor was rumored to have escaped from Nanjing before Yung-lo's troops entered the city and emigrated to some distant region. Fearing his return and potential challenge, according to this explanation, Yung-lo was eager to locate him and bring him back to China. But since no effort was made to search for the emperor during the expeditions, this explanation is the least convincing.[74]

A second explanation for the seven naval expeditions suggests that they originated from the growing power of the court eunuchs, among them Cheng Ho. The expeditions were organized allegedly to increase the prestige of the eunuchs, and of Cheng Ho in particular, by extending the emperor's influence over the maritime world, satisfying his desire for power. Moreover, Cheng Ho increased his stature in the eyes of the emperor, as well as of the court officials, by bringing back to China exotic products and animals (such as ostriches and giraffes).[75] This explanation is probably correct, but it explains only the specific event of Cheng Ho's expeditions, ignoring the geopolitical environment and the historical trajectory leading to them.

Both these explanations are based on the assumption that China was not a

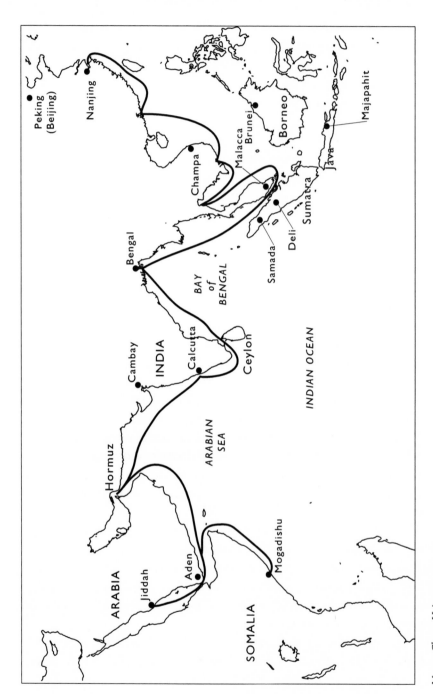

Map 3. Cheng Ho's voyages, 1405–1433

rising sea power. From this assumption, it follows that the seven expeditions of Cheng Ho were exceptional and, because of their contingent nature, strategically irrelevant. In reality, the naval expeditions of the early fifteenth century represented the pinnacle of China's growing sea power, which had expanded over the preceding centuries following clear geostrategic priorities. These priorities reflected the underlying geopolitical reality shaped by the configuration of trade routes.[76] As mentioned above, sea commerce in the South China Sea and the Indian Ocean was important for China's economy because it supplied it with key resources. The internal upheavals provoked by the Ming conquest of power and later by the civil war after Hung-Wu's death led to the slowly declining safety of sea lanes in the region. The bulk of China's power and attention was devoted to domestic conflicts, to the detriment of expansion toward the southern sea lanes. Moreover, Hung-Wu's prohibition of maritime ventures diminished Chinese influence over the South China Sea. As a result, pirates, many of them Chinese, took over several strategic locations, making travel and trade on the sea unsafe. For instance, in the 1370s a Chinese warlord established a stronghold in Palembang, in Sumatra, from where he terrorized shipping in the Malacca Strait.

Furthermore, China's absence in the region led to local powers' vying for primacy and eventually to instability. For instance, cities situated on important sea routes in Java began to wrest power from the central Javanese government, fracturing Java's hegemony in the region. The weakening of Java emboldened Sumatra to attempt to establish its own sphere of influence over the Malacca Strait, granting protection to cities such as Palembang, which rejected Javanese authority. By the early fifteenth century the region was highly volatile and descending into political chaos. The sea lanes connecting China to important markets in Southeast Asia and India, including the pivotal Malacca Strait, were seriously threatened.

In light of the geopolitical situation in the region, Cheng Ho's voyages were not mere manifestations of extravagance or curiosity. They reflected the geostrategic need to reestablish Chinese influence over these sea lanes in order to safeguard the flow of vital goods. For instance, during the first three expeditions Cheng Ho defeated the Chinese pirates infesting the Malacca Strait, destroying their fleet and strongholds.[77] Moreover, he established a base for his fleet in Malacca, which served as an entrepôt for goods. Usually, the main Chinese fleet docked in Malacca and then divided into smaller fleets that proceeded to different ports in Ceylon, India, Persian Gulf, or East Africa. On their return trip the smaller fleets deposited the acquired goods in special warehouses in Malacca before reuniting and heading back to China as one large fleet.[78] The Malacca Strait was the cornerstone of China's maritime geostrategy.

While it is unclear whether the Malacca base was garrisoned by Chinese troops between expeditions, China generally did not establish permanent colonies. "Unlike the Portuguese, who a hundred years later bulldozed their way into the South China seas by building a string of forts, the Chinese simply arranged to replace unfriendly foreign leaders in countries where they encountered difficulties with someone who was willing to trade on their terms."[79] This difference was not a reflection of Chinese moral superiority but simply of their practicality. They did not think it was efficient or feasible to defend Chinese imperial outposts hundreds, if not thousands, of miles away from the mainland. It was less costly to establish alliances in the form of a tribute system with local powers, which obtained commercial benefits when they cooperated and were castigated politically and even militarily when they did not. Often Cheng Ho's expeditions would bring back both friendly and rebellious leaders. Those that supported China were allowed to return on the succeeding expedition, while those that questioned the Ming emperor's authority were executed or imprisoned.

The last expedition occurred in 1433. After that China retreated from the sea, shut down its ports and dockyards, and destroyed Cheng Ho's maps. The sudden end of Cheng Ho's voyages is even more puzzling than their beginning. Some argue, as does Mark Elvin, that because these expeditions were "prestige ventures, . . . the lustre which they shed on the eunuch irritated many of the regular bureaucrats,"[80] and consequently it was the bureaucratic opposition of civil servants at the imperial court that forced Cheng Ho to end his trips. While the bureaucratic explanation has some validity because it points to the growing power of the civil servants in the second half of the Ming dynasty, it does not take into account the larger geopolitical situation of China.

Nor, however, does the cultural argument, which explains China's withdrawal from the sea as influenced by "strategic culture," defined as "ranked grand strategic preferences derived from central paradigmatic assumptions about the nature of conflict and the enemy, and collectively shared by decision makers."[81] It defines the role of war in human affairs, the nature of the adversary, and the efficacy of military force and applied violence.[82] In the case of China, the "cultural" explanation is grounded in the belief that Chinese international behavior was shaped by a distinct Confucian culture that abhorred the use of force to compel enemies and was wary of expansionistic goals.[83] Consequently, according to this explanation, just as the beginning of Yung-lo's sea expeditions is considered a departure from traditional Chinese foreign policy, so the end of Cheng Ho's trips is considered the return to its "normal" course. In F. W. Mote's words, "Their abandonment soon after the Yongle [Yung-lo] emperor died reflects the regularity of Chinese foreign relations."[84] Such "regularity" of China's foreign policy was defined by the Confucian beliefs shared by its ruling class. After Yung-lo's death, China returned to its Confucian roots, which had never generated an "urge for conquest."[85]

The "normal" Chinese geostrategy, therefore, focused on the territory between the steppes in the north and the ocean in the south and kept foreign relations to a minimum. When foreign contacts were unavoidable because of either geographic proximity (e.g., with Korea or the Mongols) or foreign interest in China (e.g., with Java or Japan), Peking established relations based on the understanding of Chinese cultural superiority. China was at the center of the world, the so-called Middle Kingdom, with the other, culturally inferior powers rotating around it. As John Fairbank, Edwin Reischauer, and Albert Craig observed, the Chinese "showed no sign of a feeling of cultural inferiority. Political subjugation may have been feared, but cultural conquest was unimaginable. Thus Chinese xenophobia was combined with a complete confidence in cultural superiority. China reacted not as a cultural subunit, but as a large ethnocentric universe which remained quite sure of its cultural superiority even when relatively inferior in military power to fringe elements of its universe."[86] The cultural argument, therefore, sees Yung-lo's expansionistic policy as an anomaly and China's retreat from the sea back into its continental borders as the return to normality.[87]

But the real anomaly is the end of China's maritime expansion and the resulting barricading behind a Great Wall in the north and coastal defenses in the south. As I have shown, China's interest in extending its influence over the key sea routes of Asia preceded Yung-lo's reign, climaxing in the first two decades of the fifteenth century. Cheng Ho's expeditions extended China's influence over the Indian Ocean and the Malacca Strait, marking the peak of Chinese sea power. The question, therefore, is how to explain not Cheng Ho's expeditions or their cessation but the sudden end of a long historical trajectory that brought China to a position of preeminence in Asia a few decades ahead of the European colonial thrust.[88]

There are two, complementary explanations, both Realist in nature, for China's retreat from the sea. According to one, the Ming were less Confucian in their outlook than is commonly believed. Like their European counterparts, Ming rulers were keenly aware of their geopolitical situation and pursued a geostrategy that tried to match it. Alastair Iain Johnston argues that

the arguments in favor of nonoffensive strategies were not based on a priori
Confucian-Mencian strategic cultural preference rankings or some moral-political
aversion to offensive uses of violence, but on contingent strategic arguments. . . .
Increases in offensively oriented activity tended to be associated with periods when
the Mongols were less threatening or Ming capacity to direct its resources to the
Mongol threat improved. Moreover, there is evidence to suggest a secular shift from
more offensively oriented behavior to more defensive and even accommodationist
behavior as the Ming capacity to mobilize resources against the Mongols decline.[89]

The second explanation is complementary to the "realpolitik mentality" argument and is directly related to my thesis. The end of China's sea power, and most
importantly of its attempt to control vital trade routes in Asia, was owing to its
geopolitical situation in the north. China could not afford to expend fiscal and
human resources to build oceangoing navies while its land borders were threatened. The situation on China's northern borders, as well as in Annam, in the early
fifteenth century consumed Nanjing's resources and attention, making maritime
expeditions to the Indian Ocean less important. Moreover, Yung-lo's multiple
campaigns against the Mongols in the north put an enormous strain on China's
economy. Similarly, the invasion and occupation of Annam (1407–27) diverted
military strength and imperial attention away from sea expeditions. The instability on China's land borders, coupled with the move of the capital from Nanjing to
Peking, which brought the imperial center closer to the Mongols, impeded the
projection of power far away on sea. As McNeill observed, a "more fundamental
reason for official abandonment of overseas ventures" was that a "formidable and
feared enemy existed across the land frontier," while the Chinese had no rival to
fear on the seas.[90]

The establishment of Peking as the new capital exemplified the geostrategic
priority of the northern frontier. It also had a more immediate impact on China's
sea power. Peking was situated in a barren region and depended on the south for
its economic survival. The transfer of resources from the south to the north,
therefore, become indispensable for the functioning of the imperial center. The
transportation network had to be set up in a radial fashion, with the hub, Peking,
distant from the resource-rich center of the country.[91] The Grand Canal, connecting the area of Nanjing with Peking, was restored in the second decade of the
fifteenth century. By 1416 more than 130 miles of the canal had been cleared and
thirty-eight locks had been installed.[92] Furthermore, the emperor ordered the
construction of a huge river fleet that could guarantee an uninterrupted flow of
grain to the new capital. Until 1415 grain had been delivered via both the internal

waterway and coastal ocean routes. After then, however, coastal grain shipping was prohibited, and the oceangoing ships were replaced by more than three thousand river barges.

The construction of the Grand Canal altered the naval needs of the empire. The move of the capital to Peking also led to a transfer of troops away from Nanjing, decreasing the manpower and readiness of the oceangoing fleet based there.[93] But above all, it transferred the main trade route of China from the sea to the continent, diminishing the incentives to maintain a seaborne force.[94]

Consequences of the Ming Geostrategy

The focus on the chronically unstable land borders in the north led to a gradual withdrawal from the sea. In 1432 Cheng Ho's last maritime expedition left the port of Nanjing. After its return, further expeditions were prohibited, and Peking stopped building large oceangoing vessels and concentrated instead on smaller coastal and river ships. China lost control of key sea lanes in Asia, including the Malacca Strait, and at the end of the fifteenth century withdrew from the waters immediately adjacent to its coasts. Also, China evacuated Annam in 1427, losing control over a strategic region situated astride sea and land routes. By the mid-fifteenth century China had ceased to be a sea power.[95]

The unstable land borders and the absence of control over trade routes were the result of a misguided geostrategy. F. W. Mote referred to the Great Wall as a mistake of statesmanship, but his comment can be applied to China's geostrategy toward both the north and the south. The failure to create a stable land border in the north consumed Peking's attention and resources, diverting them away from the equally important geostrategic goal of controlling trade routes. China's withdrawal from the sea led to a loss of control over key Asian trade routes connecting China with Europe through India, the Persian Gulf, and the Red Sea.

From the mid-fifteenth century on, China's geostrategy failed to reflect the underlying geopolitical reality. It did not stabilize the land borders, extend control over trade routes, or protect its center of resources. As a result, Ming China started to decline, and in 1644 the Ming dynasty was replaced by the Manchus.

The mismatch between geostrategy and geopolitics—caused by the conflicting needs to control trade routes in order to gain relative power and to defend land borders in order to avoid political annihilation—had short- and long-term consequences for China. The short-term consequences were continued instability along the land borders in the north and a growing coastal threat from pirates. The long-term consequences were China's inability to respond to the European, initially Portuguese and later Dutch, Spanish, and British, challenge in South and East Asia.

Short-Term Consequences

The short-term consequences of the failed Ming geostrategy toward China's land borders and trade routes were chronic instability in the northern region and a growing sea threat in the southern coastal areas. I have already described the situation on the northern frontiers of China, where the threat posed by unfriendly Mongol and Manchu tribes was exacerbated by an inept Ming policy that limited trade and opted for a fixed defensive posture. The Mongols continued to raid China, putting Peking under siege again in 1550. Often Chinese commanders simply refused to fight—fearing a defeat and consequent personal reprisal from the emperor—and preferred to hide inside armed camps, leaving the Mongols an open road. In 1570 a treaty gave some commercial freedom to the Mongols, abating their aggressive stance toward China.

While the Mongols were losing their aggressive fervor, the Jurchens were becoming stronger and more assertive. The peace established between the Ming and the Jurchens early in the sixteenth century on the basis of free trade ended in the 1570s, when the Jurchens, again cut off from trade and tribute, attacked China. With the aid of Chinese defectors, who supplied the financial and administrative know-how, the Jurchen leader, Nurhachi, established a government and extended his control over Manchuria and its non-Jurchen people, in 1616 assuming the title of emperor of the Chin dynasty.[96] Nurhachi died in 1626 and was succeeded by Abahai, who attacked Korea and then China. By then the relatively young Manchu state was a powerful one that considered a peace with the Ming temporary and was bent on expanding to China. As the Manchu khan argued in 1627, "If the Chinese send someone to negotiate peace with us, let us obtain gold, silver, and silk and get our escapees back. It will be a peace in name only."[97] By the 1630s, therefore, it would have been difficult for the Ming to prevent an insistently pursued Manchu expansion. Moreover, Ming static defenses proved to be as ineffective against the Manchus as against the Mongols. In the 1630s Abahai harassed northern China, and in 1644 he reached Peking unopposed. The Ming emperor, Ch'ung-chen, hanged himself, thus ending the last native Chinese dynasty. Chinese defenses in the north proved incapable of protecting China's territory in the north.

The short-term consequence of Ming maritime geostrategy was that China's withdrawal from the sea created a vacuum that was filled by smugglers and pirates, who not only cut China off from the main sea lanes of Asia but also threatened the continental heart of the empire. These pirates presented a threat akin to that of the Mongols and Manchus in the north.

China had never been threatened by a noncontiguous power. China had had peaceful relations with Japan, except when the Japanese invaded Korea in the thirteenth century. The other East Asian powers, such as Java, were too weak to present a threat, especially since until the fifteenth century China controlled the main sea lanes in the region. Ming China's withdrawal from the sea in the second half of the fifteenth century, however, left a vacuum that was gradually filled by pirates of various nationalities. In the mid-1500s a powerful group of pirates, the Wokou, probably sponsored by Japan, had become the main threat to Chinese coasts. Ming emperors succeeded in maintaining peaceful relations with Japan until 1548, when the tribute missions suddenly ended following a series of incidents. From that year on, the activities of pirates off the Chinese coast intensified, not only making sea shipping difficult but also invading the coastal areas under Ming authority. Although it is still unclear whether the pirates were Chinese renegades or Japanese adventures, the Ming court was convinced of Japan's involvement with the pirates and continued to contact Japan's authorities demanding a cessation of the plundering raids.[98]

In 1552 the Wokou pirates, allegedly fielding an army of ten thousand, invaded and established control over a whole coastal district in China. They continued to harass China's coastal areas, particularly around the Yangtze delta, penetrating as far as Nanjing, the old capital. The Yangtze region, including the Shanghai coast, was the most fertile area of China and served as its breadbasket. The wealth of the region attracted pirates, and the geographic layout made their raids relatively easy. Winds blow from Japan toward the Chinese coast in the spring and fall, and the flat coast has several natural harbors from where it is easy to penetrate deep into the internal regions through navigable rivers.

The ease with which the pirates extended their control over the sea and the coastal territories of China served as a wake-up call for the Ming court. Over the thirteenth and fourteenth centuries, coastal defenses had been neglected. Cities had no walls and were effortlessly conquered by the Wokous; Ming troops were poorly equipped and led by incompetent officers; and most importantly, there was no Chinese navy to prevent the Wokou sea invasion.[99]

Ming China had lost its naval skills and responded "unimaginatively and ineffectively with coastal defense measures."[100] The Ming administration tried to develop adequate defenses by walling the cities, rebuilding some of the fleet, and strengthening the military garrisons along the coast. The strategy was similar to that implemented in the north against the steppe nomads: purely defensive, positional, founded on the belief that walls could prevent the invasions of a highly mobile enemy. And it was equally ineffective in protecting China's coasts from the plundering raids of the Wokous.

To defeat the Wokous it would have been necessary to deprive them of their offshore launching bases. But as Charles Hucker observed, this

> would have required Chinese conquest and control of the Liu-ch'iu Islands, Taiwan, and even part of Japan itself. Early in the Ming dynasty, when the famous eunuch admiral Cheng Ho led huge armadas across the Indian Ocean, the establishing of an overseas Chinese empire might have been possible. Both T'ai-tsu (1368–1398) and Ch'end-tsu (1420–1424) at least tried to intimidate Japan with threats of invasion. But even these early Ming emperors were not foolhardy enough to carry out such threats, and later emperors had been so concerned about possible troubles in the north that they had allowed Chinese seapower to decline; no one could have suggested defense-by-conquest for serious consideration in the 1550's.[101]

The threat of pirates had effectively ended by the 1570s. Even though smugglers continued to proliferate off the Chinese coasts, there were no large-scale invasions by sea after the 1580s. In part the abatement of the Wokou threat was due to Ming efforts. Peking devoted enormous resources to combating the coastal pirates by erecting walls around the cities of Shanghai and Nanjing and by sending its best troops and generals to the region.

But it took twenty years for the Ming to eradicate the Wokou threat. Consequently, it is doubtful whether China alone could have completely defeated the pirates because it had minimal control over the sea and the offshore bases of the pirates. To Ming officials, "the sea represented problems, not opportunities, and statecraft stopped, if not at the water's edge, certainly short of high seas."[102] In fact, it is probably more correct to argue that the decline of the Wokou threat occurred in part because of the entry of European powers, into Asia, in the sixteenth century the Portuguese empire in particular. The arrival of the Portuguese filled the vacuum left by China's withdrawal from the sea, depriving the pirates and smugglers of the freedom to roam the sea in search of plunder. Once Portugal had stabilized the region and established itself as the main intermediary in the trade between China and Japan, "the illicit smuggling activities in the Chekiang and Fukien waters, which had invited the imperial wrath, were no longer attractive. The Portuguese much preferred this newly won status [of commercial brokers with Japan] and were willing to lend the [Chinese] authorities a hand in the suppression of the smuggling-piratical activities."[103]

Long-Term Consequences

For China the benefits of Portuguese help against the pirates were quickly outweighed by negative consequences. Having turned inward in the fifteenth century, China was poorly equipped to challenge the growing European presence in Asia, and it faced the Portuguese colonial powers, and later the Dutch and the Spanish, from a position of weakness. Despite the failure of Europeans to penetrate deeply the Chinese mainland, Peking could not prevent the colonization of its shores, starting in Macao.

In the sixteenth century European powers, led by Portugal, rapidly extended their control over the key sea lanes in Asia, from the Indian Ocean through the Malacca Strait to the South China Sea. The Portuguese built their sea empire on several strategic bases along the sea lanes connecting Europe to China. Led by Afonso de Albuquerque, the Portuguese conquered Goa, on the western coast of India, in 1510 and Malacca in 1511.[104] By establishing control over Hormuz, Goa, and Malacca, Albuquerque threatened the Arab monopoly over the spice trade in the Persian Gulf, the Indian Ocean, and the Indonesian archipelago. Arab ships, in fact, carried spices and silk from the Malacca Strait (where Chinese merchants delivered their goods until the late fifteenth century) to India, the Persian Gulf, and the Red Sea, and from there to the Mediterranean Sea through Ottoman lands. It is true that the Portuguese failed to close the Red Sea, anticipated by Ottoman expansion to Aden, and Asian spices and goods continued to trickle to the Mediterranean through this route. Nonetheless, the constant Portuguese harassment of Arab shipping in the Indian Ocean increased spice prices in Alexandria, undermining the Ottoman and Venetian monopoly over this trade in the Mediterranean.[105]

The next step for the Portuguese was to extend their control past Malacca and the Indonesian archipelago.[106] In 1514 the first Portuguese ship reached China, and three years later a few ships reached the port of Canton, where the arrogant and rambunctious behavior of their sailors provoked a small incident with the Chinese authorities. The relations between the Portuguese and the Ming continued with ups and downs during the following two decades. As a result of quarrels between the Portuguese and the local Chinese authorities, in 1522 the Ming defeated and expelled the Portuguese from Canton, forbidding trade with them. But in the mid-sixteenth century, probably merely approving an already sizeable colony of Portuguese merchants, the Chinese granted Portugal the port of Macao, in the south of China, which remained a Portuguese colony for more than four centuries.

By granting ports as well as de facto monopoly over China's foreign trade Peking tried to control the Portuguese onslaught in Asia.[107] For instance, the gift of Macao to the Portuguese was a shrewd move that allowed Peking to manage Portuguese influence in China. China could in fact easily hold sway over the activities of Macao because it was located close enough to China's markets to allow a profitable trade and because it was connected to the mainland only by a narrow stretch of territory that was under Chinese control. Thus, the Ming could regulate Portuguese access to the mainland markets, maintaining some leverage over the European power.[108] The gift of Macao, therefore, was a sign of China's weakness, not of its superior generosity.

Moreover, China was willing to give Portugal the monopoly over its maritime trade because of its inability to pursue it alone. In the mid-sixteenth century the Chinese were wary of trading with Japan, partly because of the Wokou threat and partly because of their own inability to control maritime shipping. Hence, rather than sending their own subjects to Japan, Ming rulers were happy to let the Portuguese serve as intermediaries. The alternative for Peking was inconceivable: to maintain commercial relations with Japan and Southeast Asia, China would have had to revitalize its maritime geostrategy, rebuild an oceangoing navy, and reestablish influence over the sea, which was already firmly under Portuguese control. Such concessions, however, further weakened Ming maritime power. From about 1570 to the end of the Ming dynasty in 1638 the Portuguese carried a lucrative trade between China and Japan, while Ming ships were rotting in ports.[109]

The role played by the Portuguese in Macao exemplified well the position attained by the European powers in Asia. The Portuguese in China did not trade European goods, but almost exclusively Chinese silk, which came from the internal markets of the mainland and was shipped to Japan. Macao was the entrepôt for this regional commerce, and its economic fortunes depended on the regional inter-Asian trade more than on its connection with Europe.[110] Spices, silk, and other exotic goods continued to bring profits back in Europe, but greater economic and political benefits were derived from controlling trade between Asian countries. In fact, the regional trade "was far more extensive and probably more profitable than the trade of the royal fleets which it helped to finance. Ironically these fleets, connecting Goa with Lisbon, formed the weakest link in the chain of the empire."[111] Control over regional trade was maintained by the Dutch, the Spanish, and the English, who challenged and gradually replaced the Portuguese starting in the last years of the sixteenth century. These European powers became the main regional actors in Asia, replacing China.

In the late sixteenth and early seventeenth centuries the main struggle in Southeast Asia and off China's shores was between the Portuguese and Dutch powers. The Dutch arrived in Asia in the late sixteenth century via Siam and Indonesia. In the early seventeenth century they became interested in the East Asian trade, especially between China and Japan and between China and Manila. Dutch ships began raiding China's coast and in 1622 attacked Portuguese Macao. Despite their strength, the Dutch were forced to settle in Taiwan, where they remained until 1662.[112] The striking aspect of the Dutch arrival off China's coast was that their enemies were either European (Portuguese and Spanish) or local warlords, pirates, or merchants, and not Chinese authorities.[113]

During the seventeenth century the intra-European competition in Asia continued on the two main oceanic routes: through the Indian Ocean under Dutch influence, which was firmly established in Malacca, and through the Pacific to America under Spanish control. The Portuguese, heavily undermanned and outgunned, succeeded in maintaining an important commercial presence (including Macao) by using the transpacific route until the mid-1600s.

China played no role in the sixteenth- and seventeenth-century struggle for hegemony in Asia. The Ming court, embroiled in futile resistance against the growing Manchu threat in the north, did not pose much resistance to European powers. The Ming had no maritime geostrategy, and the struggle in Asia became a struggle between European empires, with China as a prize but not an actor.

China's absence in the sixteenth-century struggle for Asia is striking given its technological development. China was technologically equal, if not superior, to European powers. And it was definitely more developed than the tribes encountered by the Spaniards in America at roughly the same time that the Portuguese were sailing into the South China Sea. From a technological point of view, Portugal in the sixteenth century and the Dutch in the seventeenth faced a much more difficult environment in Asia than Spain faced in the New World. Not only Arab pirates but also the weak Chinese coastal fleet managed to inflict a few defeats on Portugal's ships in the second half of the sixteenth century.[114] China in particular could have posed a serious challenge to the Portuguese because of its high level of technological development. For example, Chinese junks were large enough to carry artillery and could withstand rough open seas and were well suited to fight against the sturdy oceangoing Portuguese ships.

Yet, by the time the Portuguese reached China's coasts, the Ming court had no intention of expending resources to upgrade its fleet. As Carlo Cipolla writes, the Chinese junk "was never developed into a man-of-war. Like the Mediterranean galley, the Chinese war-junk remained essentially a vessel suited for ramming and boarding. With very high castles and no portholes for guns in the hull, the war-junk was fit only for the traditional way of fighting and such it remained."[115] The tactical differences in naval warfare decided the equilibrium of power on the Asian seas. The encounter between Chinese and European vessels was one between two fundamentally different ways of fighting battles on sea. The Chinese naval tactics were similar to those of the Ottomans and the Venetians, relying on close combat, which effectively transformed a naval battle into a land one.[116] The Portuguese instead had adopted artillery as the main offensive naval weapon, which allowed their ships to hit the enemy's vessel from a safe distance. With their superior artillery and tactics, the Portuguese navy defeated the Ottoman fleet in the Indian Ocean in the early 1500s. By then the Chinese fleet did not possess the artillery and skills needed to compete with the European fleets. But we may wonder what would have happened had the Ming continued their early involvement in maritime affairs. As one historian put it, "If the Ming state had not abandoned its great maritime venture, the Portuguese would have found it much more difficult to get a foothold on the coast of India, and probably could have accomplished nothing in Melaka, Sumatra, and Siam."[117]

It has been argued that there was no connection between European control of the Asian sea lanes and the fate of Ming China. As John Wills put it, "Because China was such a great continental mass, maritime power could not be a determining factor in its political destiny."[118] Because Ming China did not pursue an expansionistic maritime geostrategy, it became increasingly isolated from foreign influences, and this isolation made it less vulnerable to European control over trade. According to this argument, foreign trade accounted for a minimal percentage of China's overall economy, perhaps as low as 1 percent in the early seventeenth century. Consequently, fluctuations in commercial exchanges controlled by Portuguese and Spanish ships (such as the drop in the supply of silver from the American mines in Spanish Potosi and Mexico in the 1630s) could not have affected the Ming decline in the 1630s.[119] Therefore, the source of the Ming decline was not external (dependence on European-controlled trade) but domestic (a weakened central authority).

The appearance of Chinese imperviousness to foreign influences was strengthened by the fact that China's mainland remained inviolate throughout the sixteenth century and the first half of the seventeenth. Portugal, as well as succeeding European powers, held an advantage over China on sea but not on land. Arguably, Chinese armies were superior to European ones, having a considerable advantage in manpower and good artillery skills. Chinese military superiority deterred the Europeans from invading the mainland. Although individual European merchants and groups of missionaries, mostly Jesuits, traveled throughout China and often settled in Peking or in other major urban centers, there had never been an attempt to penetrate deep inside China's mainland.[120] Portugal, like the succeeding European colonial powers, obtained control only over ports and limited coastal areas.

In reality, China was affected very negatively by European control of the Asian sea lanes, and the preservation of its territorial sovereignty was deceiving. First, the fact that China maintained its territorial integrity despite the European onslaught in Asia does not mean that Peking continued to play an important role in the region. European powers simply did not need to control China's mainland in order to become regional great powers. Portugal and its European competitors achieved influence and power in East Asia without controlling China's mainland and therefore without fielding enormous armies, which would have been logistically difficult to achieve and too costly to sustain for a protracted period of time. The Portuguese controlled a few fortresses in strategic locations (e.g. Hormuz, Goa, Malacca, and Macao) and based their power on them. Albuquerque, the fifteenth-century Portuguese conqueror, argued that the entire Portuguese empire in Asia could have been defended "with four good fortresses and a large well-armed fleet manned by 3,000 European-born Portuguese."[121] The four strongholds were Hormuz, which closed the Persian Gulf; Goa, which controlled the Indian Ocean; Malacca, which commanded the passage from East Asia to the Indian Ocean; and Macao, which ruled over the South and East China seas and had access to China's internal market. From the early sixteenth century to the first decades of the seventeenth, despite its chronic lack of ships and manpower, Portugal maintained a tight control over the sea lanes of the region. According to some estimates, no more than ten thousand Portuguese men protected the trade routes from East Africa to Macao.[122]

Second, the lack of Chinese sea power and of control over the main Asian trade routes had a negative impact on the fate of Ming China. The mastery over the main sea lanes in Asia allowed Portugal, and later in the seventeenth century the Netherlands and Spain, to control commercial relations among Asian countries, achieving enormous political leverage over the region.[123] More precisely, European influence over Asia, and China in particular, can be divided into direct and indirect. The direct influence consisted in the ability of European powers to dictate terms of trade to China and other Asian countries. "Native" Asian shipping was never extirpated and replaced by European fleets because of the technical difficulties associated with building and manning navies large enough to accommodate the trade flow. Chinese junks, manned by Chinese merchants, continued to set sail from the Fukien ports throughout the sixteenth and seventeenth centuries. But their routes, and above all the composition and price of their cargoes, were subject to Portuguese, Dutch, or Spanish approval. European powers controlled the key passages, where they demanded fees and taxes.[124] Ships that did not possess a permit from the appropriate authorities were subject to confiscation or destruction.[125] As Cipolla writes, "Maps show better than any verbal description that until the eighteenth century European possessions the world over consisted mostly of naval bases and coastal strongholds. . . . Within a few years after the arrival of the first European vessels in the Indian Ocean it became mandatory for non-European vessels to secure sailing permits if they did not want to be blown up by European guns. The oceans belonged to Europe."[126]

The lack of Chinese sea power and the resulting lack of control over vital trade routes also had a profound indirect impact on China's power. Despite the assertions of some historians, mentioned above, it is still unclear how exposed the Ming economy was to fluctuations in trade flows and to external markets. But there is some consensus that the trade silver and luxury goods such as spices collapsed in the early seventeenth century.[127] This commercial slump was not owing to China's internal problems but to external factors. For instance, in the 1630s silver mining in Mexico and Potosi in the New World decreased dramatically. Because silver was in high demand in China, the state of silver mining in Spanish America had an impact on the Chinese economy, driving prices up.[128] China's welfare and power depended on the situation in America, over which it had absolutely no control. Hence, "Sino-Spanish trade in Manila decline after 1635 not because Chinese goods were unavailable but rather because the Spanish did not receive sufficient silver from the New World to pay for them."[129]

Peking had no independent ability to bring more silver or to seek other markets for its silk because the trade was entirely in European hands.[130] China was at the mercy of the fortunes of European empires. Goods flowed to and from China only as long as European navies were able and willing to carry them.

Intra-European conflicts also affected Chinese merchants' ability to obtain resources and sell their goods. In particular, the Dutch-Portuguese struggle for mastery of the Asian seas had a negative impact on China's power. Dutch ships disturbed Portuguese shipping in the South China Sea beginning in the early seventeenth century. After the Dutch conquered Malacca in 1641, Portuguese ships based in Macao were cut off from the Indian Ocean, limiting China's indirect access to European markets.[131] Again, by then the Ming court had no means of opening a new route to access foreign markets. The failure to continue an active maritime geostrategy after the second half of the fifteenth century became painfully tangible in the sixteenth and early seventeenth centuries. China's lack of interest in maritime activities and its inability to control them undermined its position of power and preeminence in Asia.

Conclusion

Ming geostrategy succeeded in maintaining and extending China's influence in Asia when it effectively protected its land borders and upheld its control over key sea lanes. The decline of the Ming dynasty began when its geostrategy focused exclusively on the continental borders, locking China in a defensive posture that failed to defeat or secure the northern frontier.[132] Such a one-sided geostrategy, initiated in the mid-fifteenth century, did not completely reflect the underlying geopolitical situation. The main trade routes fell outside of China's control, and by the early sixteenth century they had become increasingly dominated by Portuguese and Dutch merchants, who fought against each other in the succeeding centuries. At the same time, China's land borders remained unstable, eventually leading to the end of the Ming dynasty, defeated by the Manchus in 1644. And the resource-rich Yangtze region remained perilously unprotected, as was made painfully clear by the Wokou threat in the 1570s.

Locked in a constant conflict in the north and unable to extend its influence over the trade routes in the south, Ming China declined slowly throughout the sixteenth and seventeenth centuries. It did not lose its territorial sovereignty, but as with the Venetian and Ottoman empires, its power was curtailed, and it never again achieved the splendor it had had in the fourteenth and fifteenth centuries. Not even the infusion of new energy that came with the Manchu conquest of Peking in 1644 helped to restore China to a position of regional power.

Was the decline of Ming power in the short term and of China in the long term irreversible? The temptation of geopolitical studies has always been to find a determining power in geography. Some have argued that Ming China was doomed from the beginning because of a geographic environment that hindered the development of sea power—that Ming China was simply geographically incapable of creating a maritime power that could counterbalance the Portuguese, Dutch, Spanish, and, later, British colonial empires. According to this argument, the geography of Asia, with its distant shores, vast open seas, and insidious current and wind patterns, did not encourage maritime contacts like those that flourished in the Mediterranean, where "power centers facing narrow and relatively safe seas sought to profit from their relations with each other through trading, piracy, plunder, raiding to force better conditions of trade, and colonization. Naval power could be used to concentrate wealth in one center at the expense of another, and the wealth would pay for the fleets. Concentration of power facilitated control of trade routes and brought the increased profits of trade monopolies."[133]

China instead faced an open ocean that not only was perilous to navigate but also, because of the distance between shores, did not bring as much profit from trade as the Mediterranean. Consequently, there was no connection between profit and power. In other words, not only was the control of sea lanes difficult but it was not crucial to achieving power in the region.

There are two problems with this argument. First, China was challenged not by Mediterranean powers such as Venice or the Ottoman Empire but by Atlantic states, Portugal being the pathbreaker. The Atlantic empires arguably faced a more difficult maritime environment than did China, and yet they extended their influence over the sea lanes of Asia, controlling regional trade and achieving strategic superiority over the local powers. China could have been the Portugal of Asia. It had the resources and the technology to extend its influence over Southeast Asia by establishing bases in strategic choke points, but the Ming rulers chose not to do so.[134]

The second problem with the argument that geography determined the Ming maritime geostrategy (or lack thereof) is that it ignores China's long past as a sea power. As I have shown, from the Song until the mid-Ming dynasty China was a maritime power that established some form of sea empire. It is true that unlike the European empires, both Mediterranean and Atlantic, China did not found state-sponsored colonies. Even an important port such as Malacca was probably left without a standing Chinese garrison. Yet, China controlled the sea lanes up to the Malacca Strait thanks to its merchants and to its naval expeditions. The history of Chinese maritime power contradicts the argument that geography determined the superiority of European sea empires in Asia.

The geographic environment did not doom China. Nor did geopolitics. The underlying geopolitical situation did not change drastically from the fourteenth to the seventeenth century. The most momentous change brought about by the European discovery of the Cape of Good Hope route was the introduction into Asia of a new player—Portugal, followed by other Atlantic European powers—and the diminished importance of the land routes in Persia. But the discovery did not alter the trade routes in Asia, which continued to go from China through the South China Sea, the Malacca Strait, and the Indian Ocean toward Europe. Similarly, the continental borders of China continued to present the same challenges from the early to the late Ming period, with the Mongols and then the Manchus pressing on the frontiers.[135]

The key change from the fourteenth and the seventeenth century was in the Ming geostrategy. After Yung-lo, in the mid-fifteenth century Ming emperors abandoned an active policy against the Mongols that could have stabilized the northern land border and at the same time ceased to maintain their influence over the sea lanes in Southeast Asia. The inability of the later Ming rulers to protect their land borders, exacerbated by the move of the capital to Peking, prevented them from concentrating their efforts on the sea, and this, in the end, led to China's fall from its position of power.

CHAPTER SEVEN

Lessons for the United States

Venice, the Ottoman Empire, and Ming China became great powers and maintained their position by exercising control over key centers of resources and the routes linking them. When their control weakened, or when they controlled routes that were no longer strategic, these great powers lost their position of primacy. The question this chapter examines is whether these historical cases offer lessons that are still relevant. Specifically, is it still relevant to talk about geography in an age of instant communication and economic globalization? And does the United States (or any other great power in the modern age) have to pursue a foreign policy that reflects the underlying geopolitics and strive to control routes and resources in order to maintain its power? In light of its superior power, modern technology, and access to the global market, can it instead ignore the geopolitical situation and devote its political attention and military resources to other, less geopolitically driven objectives? Finally, what lessons can the United States learn from the three historical cases examined in the previous chapters?

Is It Still Relevant to Talk about Geography?

The policy relevance of this book is predicated on the belief that geography, and specifically the distribution of resources and the arrangement of routes, continues to matter for foreign policy. As I pointed out in chapter 1, the study of geography is not as popular today as it was fifty years ago, but its ups and downs are tied more to academic mood swings than to fundamental changes in the way states interact. In fact, geopolitics, and consequently the need to pursue a geostrategy that reflects it, continues to be important even in the age of instant communication and "globalization." Control over resources (such as oil) and access to routes (e.g., sea lanes, which are vital for the transport of oil as well as for projecting military power) continue to be key strategic objective of states.

There are, of course, many arguments against the importance of geography.

Most stem from the belief that over the past decades there has been a radical change in international relations caused by new technologies (e.g., nuclear weapons and especially missile technology) or new economic structures (e.g., based on services and knowledge, rather than raw materials or land). It is argued that such changes make the geographic distribution of resources and the routes linking them less important (e.g., in the case of "knowledge economies") and that states should thus take less interest in geopolitics and in pursuing policies that reflect it. Perhaps the most persuasive of these arguments falls under the broad category "economic globalization." This argument is based primarily on a belief in the economic incentives of a free global market. The existence of such a market diminishes the incentives to exercise direct control over natural and economic resources. It is easier to buy them on the market than to expend power to control them. In other words, the prerequisite of power is not control over resources but the ability to purchase them. Hence, the difference between, for instance, the Ottoman Empire and a modern great power is that the sultans had to maintain control over the key routes and centers of resources in Eurasia in order to preserve their power, while a twenty-first-century state can simply buy key resources on the market. A global market makes traditional geostrategic imperatives obsolete.[1]

Such an argument, however, is at best incomplete. A globalized economy does not diminish the importance of geography, for three reasons. First, economic interdependence increases the ties binding a state to distant regions (e.g., China to the oil-rich Middle East, or the U.S. economy to that of Europe or Asia). As a result, states are forced to pay greater attention to such regions because what happens there affects their economic and political fortunes. Because of limited resources, states have to discriminate according to some geographic criteria and focus their diplomatic and military energies on some regions, while avoiding others. Some areas of the world (e.g., Africa) are not centers of economic or natural resources, and consequently their fate is less vital to the security and welfare of the United States and other powers. Geostrategic prioritization is as vital for today's great powers as it was for Venice or Ming China.[2]

Second, globalization does not mean that trade occurs in cyberspace. In other words, globalization occurs through channels—land routes, sea lanes, air corridors—that are defined by geography. For instance, 90 percent of Japan's energy supply is delivered via the South China Sea. Similarly, 40 percent of U.S. trade is transported via sea lanes.[3] The stability of these sea lanes is vital to the economies of the United States and states in Asia and Europe. Therefore, it is imperative for states to monitor and, if necessary, protect and patrol these routes, as was done during the times of Venice and Ming China. Globalization, in other words, has not detached the interactions of states from geography.

Finally, the argument that globalization frees states from traditional geostrategic concerns ignores the role of the United States in the world. To use again the example of maritime trade, the current "freedom" of sea lanes is a result, not of globalization, but of the American preponderance of power, which keeps trade routes open to global traffic, thereby allowing U.S. allies to ignore geopolitics and the need to pursue a foreign policy directed at protecting and controlling key trade routes. For instance, Europe and Japan do not have to project power to the Persian Gulf or the South China Sea because the United States already does that, ensuring the free flow of products and resources to the markets. As Barry Posen observes, control over these routes, which he calls the "commons," is firmly in U.S. hands even though they,

> in the case of the sea and space, are areas that belong to no one state and that provide access to much of the globe. Airspace does technically belong to the countries below it, but there are few countries that can deny their airspace above 15,000 feet to U.S. warplanes. Command does not mean that other states cannot use the commons in peacetime. Nor does it mean that others cannot acquire military assets that can move through or even exploit them when unhindered by the United States. Command means that the United States gets vastly more military use out of the sea, space, and air than do others; that it can credibly threaten to deny their use to others; and that others would lose a military contest for the commons if they attempted to deny them to the United States.[4]

But such a preponderance does not mean that geopolitics has ceased to be a source of contention and strategic fears. U.S. allies, such as Japan in the Pacific, must be prepared for a U.S. strategic retrenchment. The World Trade Organization be a powerful organization, but it will not guarantee the free flow of oil through the Strait of Hormuz or the commercial exchanges between Europe and North America or between China and the rest of Asia. Moreover, potential challengers (e.g., China) resent having to rely on the United States to protect their sources of energy or trade. Such protection is a form of leverage, especially in case of a conflict, when the United States could cut its enemies off from their access to resources and markets.

The distribution of resources and the configuration of trade routes—in brief, geopolitics—will therefore continue to influence international relations. The geopolitical situation of the world has not been overcome by new technologies or economic structures. Consequently, it will continue to be necessary to incorporate geographic factors in the foreign policy of states: some regions will be more strategic than others and as a result will demand greater attention.

Is Geography Relevant for the United States?

The United States, like any other great power in history, has to pay close attention to geopolitics and adapt its geostrategy to reflect geopolitical changes. However, as in the case of geography, there is a temptation to discount the constraints geopolitics imposes on U.S. policy. Specifically, the main challenge is to recognize the limits of U.S. power and its inability to change or overcome geopolitics. As described in chapter 2, geopolitics, the layout of routes and resources, is not under the individual control of states, which, no matter how powerful, have to accommodate their geostrategy to the underlying geopolitical reality. At the peak of their power the Venetians, the Ottomans, and the Ming emperors forgot for a variety of reasons that their success as the most powerful states of their time depended on the constant monitoring of the geopolitical situation and on the reflection of this situation in the geostrategies they pursued. In the United States today there is a similar feeling that power, both military and economic, can trump geography and make the geographic prioritization of strategic objectives unnecessary. In other words, geostrategy can turn into a policy of global aspirations, with no, or at best limited, geographic discrimination between regions. In part, such temptation may be owing to the position of supremacy currently enjoyed by the United States, which makes global influence and hence global aspirations possible. But in part it may be owing to a more deeply entrenched faith in human capabilities to overcome limits. As Reinhold Niebuhr put it, writing on the United States, "It is particularly difficult for nations to discern the limits of human striving and especially difficult for a nation which is not accustomed to the frustrations of history to achieve this moderation. It is not sloth or the failure to exploit our potentialities but undue self-assurance which tempts the strong, particularly those who are both young and strong."[5]

However, it is important to remember that as in the case of past empires and great powers, so in the case of the United States the understanding of geopolitics and the necessity to reflect it in foreign policy remain vital. The study of geopolitics is especially crucial for the United States. Over the past decade the United States has faced, and in coming years will continue to face, a geopolitical change of proportions similar to those faced by Venice, the Ottoman Empire, and Ming China. The rise of China as a great power, combined with the decline of the Soviet Union (and Russia) and the fraying of the transatlantic relationship, will result in a geopolitical situation that is profoundly different from that of the previous century. This will demand a reevaluation of geostrategic priorities, including the forming of new alliances and the development of new technologies. Heeding geopolitics is also vital for the United States because of its reliance on routes, especially sea lanes, for economic and military activities. Control over these routes is critical for the preservation of U.S. economic growth and political influence in the decades to come.

The United States is facing a geopolitical change similar in scope to those faced by the three great powers examined in this book. The rise of China as a new great power in Asia, combined with a weak and chaotic "Heartland" in Eurasia (Russia and the so-called arc of instability, from the Black Sea to Central Asia) and a potentially unfriendly, or at least unsupportive, Europe, will lead to a new geopolitical situation, dramatically different from the one that characterized the twentieth century.[6] As a 2004 report by the National Intelligence Council states,

> The likely emergence of China and India as new major global players—similar to the rise of Germany in the 19th century and the United States in the early 20th century—will transform the geopolitical landscape, with impacts potentially as dramatic as those of the previous two centuries. In the same way that commentators refer to the 1900s as the "American Century," the early 21st century may be seen as the time when some in the developing world, led by China and India, come into their own.[7]

The most evident change would be the presence of a great power, potentially hostile to the United States, on the shores of East Asia, facing the Pacific Ocean. This would represent a significant shift from the threats previously faced by the United States. Since the late nineteenth century, with the rise of Germany and later of the Soviet Union, the center of power in Eurasia has been continental, and U.S. foreign policy has focused on circumscribing the reach of the land power. The U.S. geostrategy has focused on strengthening, through alliances, the U.S. presence in the neighborhood of the threatening land power and maintaining a robust naval force, especially in the realm of antisubmarine warfare, to limit the enemy's maritime reach.

However, the new challenge will be from China, a power that has easy access to oceans and the interest and the capacity to develop a blue-water force sufficient to threaten American maritime hegemony in the Pacific.[8] In fact, Beijing has strong maritime interests because its principal strategic objective, Taiwan, requires the development of naval capabilities. Moreover, China became a net oil importer in 1993, and its energy needs continue to grow rapidly. As a result, Beijing is becoming increasingly more dependent on the supply of oil from the Middle East, which, despite China's attempts at diversification (especially through agreements with Central Asian republics and Russia), is its principal supplier.[9] In other words, China depends on the sea for its economic and strategic needs. The challenge is that the United States, not China, controls these Asian sea lanes. American maritime supremacy guarantees the free flow of oil to Japan as well as to China, but it is a source of enormous leverage for the United States in case of a conflict with Beijing. If tensions between the two powers increase, the United States may decide to limit, and even cut, the flow of oil to the Chinese economy. To use a historical example from this book, the state that controls these Asian sea lanes is in a position similar to that of Venice, which almost monopolized the key routes linking Asia and Europe several centuries ago. The other states, from France to the Papal States, were forced to rely on Venetian power to achieve their objectives (e.g., the Fourth Crusade in 1204), sacrificing in the process their strategic independence. To prevent a similar situation, which would constrain China's ability to pursue its strategic objectives (e.g., Taiwan), Beijing has been modernizing its military, especially its navy, with the aim of developing capabilities, such as submarines and missile technology, that would threaten the U.S. command of the seas.[10]

Obviously, the key question whether China will expand cannot be answered by studying geopolitics because geopolitics does not determine states' courses of action. Venice was not forced to expand toward the eastern Mediterranean, and Ming China's expansion toward the sea (and its withdrawal from it) was not determined by the layout of resources and routes. However, the geopolitical situation can suggest where a state is likely to expand if it chooses to increase its power. If China decides that it wants to be a great power with global or regional aspirations rather than a status-quo power concentrated on its domestic issues, it will focus on the control of the sea lanes in the Pacific.

Moreover, as the case studies illustrate, territorial security, or stable borders, is a precondition of any expansionistic foreign policy. The instability on their land borders hampered Venice's and Ming China's efforts to project power to, respectively, the eastern Mediterranean and the South China Sea. The trade-off between defending territorial safety and projecting power to distant regions continues to apply today. Modern China has traditionally been too preoccupied with its long land borders to develop a navy powerful enough to become an actor in the Pacific Ocean. However, the weakening and then the collapse of the Soviet Union altered this geopolitical situation faced by China. Since 1991 Beijing has pursued a skillful diplomacy aimed at stabilizing its relations with Russia and with the new

Central Asian states. In 1991 it reached an agreement with the then Soviet Union demarcating about 98 percent of their common border, erasing some potential sources of conflict, and by 2005 only one border dispute (with India) remained unresolved.[11] In 1996–97 Beijing signed military agreements with China, Russia, Kazakhstan, Kyrghistan, Tajikistan that Russian officials described as "non-aggression treaties." The stabilization of China's land borders may be one of the most important geopolitical changes in Asia of the past few decades. From a tense frontier similar to that of Ming China it is turning into a stable one that does not require an enormous expenditure of military strength or political attention.[12] This might free China from having to devote resources and attention to its land borders, allowing it to pursue a more aggressive maritime geostrategy.

China's steady accumulation of power and the resulting change in the geopolitical situation will force a reevaluation of U.S. geostrategy. The main theater of action will shift from Europe and the Atlantic to East Asia and the Pacific, requiring the United States to develop new technologies and alliances. The fundamental strategic objective of the United States will change from containing a predominantly continental power (the Soviet Union) to limiting the maritime reach of a budding sea power (China). In the case of the Soviet Union the United States had to maintain a strategic foothold in the Eurasian littoral, Western Europe, and Japan to put pressure on Soviet territory and maintain the U.S. maritime link with these allies.[13] The goal was to prevent or at least blunt a Soviet land assault in case of a violent confrontation. In the case of China, the U.S. objective is to prevent China's maritime expansion, not to defend a continental outpost in Asia. The United States will have to rely on its sea power to maintain its position of supremacy in East Asia, and for this purpose, unlike in the European theater, it can use only a handful of bases (e.g., Japan, Diego Garcia [Indian Ocean], South Korea).[14] The challenges presented by the geography of the new theater were made quite clear during the 2001–2 war in Afghanistan. The lack of established ties with many of the states in Central Asia forced Washington to create ad hoc alliances enabling it to use military bases for the war. The lack of bases in the region made it difficult to use efficiently many tools of the American arsenal, notably its technologically superior but short-range fighter planes. They were built with the European theater—characterized by short distances, a very hostile environment, and a large number of closely spaced bases—in mind. The Pacific presents fundamentally different challenges, such as long distances, few bases, limited U.S. ground operations, and potentially lethal Chinese anti-ship capabilities.

The recognition of a geopolitical change, therefore, needs to be followed by a corresponding adjustment in geostrategy. Geopolitical changes create serious challenges but are rarely the ultimate determinants of the fate of states, especially of great powers, which tend to have the necessary strategic flexibility and power to adapt.[15] This is not to say that changes in geostrategy are easy to formulate and implement. On the contrary, as the historical cases examined earlier demonstrate, it is very difficult to adjust a geostrategy to a new geopolitical reality, and often the end result is a disconnect between geostrategy and geopolitics. Venice and the Ottoman Empire, for instance, faced geopolitical changes that altered dramatically their positions of power and had limited strategic responses to choose from. Ultimately they failed to adapt to the new geopolitical situation, even though both, but especially the Ottomans, expended a great deal of effort to rearrange their geostrategies. But their failure should not lead us to believe that great powers are doomed to decline when there is a shift in the distribution of power and the pattern of routes. Decline is not inevitable. In the case of the United States facing a new geopolitical situation, how the United States adapts its military power and its alliances to the new theater, East Asia and the Pacific, will determine whether American supremacy can last in the twenty-first century.

It is also important to consider geography, and geopolitics in particular, because the United States is heavily dependent on resources from abroad and on the routes linking the American homeland with the world. In fact, because of the rapid growth of international trade (between 1997 and 2000 U.S. international trade increased by more than 28 percent), the United States relies increasingly on the flow of goods from specific regions of the world and consequently needs to pay close attention to the geopolitical situation in those regions.

First and foremost, the bulk of U.S. trade crosses the seas, making the United States dependent on the oceans.[16] Obviously, the observation that the United States is a maritime power is not new. Admiral A. T. Mahan spent much of his intellectual career at the turn of the twentieth century alerting the American public and policymakers to the importance of sea power for the United States, especially in light of the building of the Panama Canal.[17] Mahan pointed out that the sea offered both an opportunity and a risk to the United States. The opportunity was the ability to project power to distant locations without the immediate threat of an irascible neighbor. The risk was that others could do the same, directly threatening U.S. survival, as well as block American commercial relations with Asia or Europe. In other words, the opportunity presented by the seas required constant protection, and the difficulty of exercising control over blue waters did not mean that maritime navigation would always be open, or "free."

Control of the seas is even more important to the United States now than during Mahan's years. For instance, by value about 40 percent of U.S. trade (40.2 percent in 1997, 38.4 percent in 2001) is seaborne. By comparison, about 27 percent is by air. If we use weight as a measure, more than 70 percent of U.S. trade is by sea (73.3 percent in 1997, 77.7 percent in 2001), while only 0.4 percent is by air. Moreover, by value almost half of U.S. imports (46 percent) come by water, compared with about 30 percent by air.[18] These data indicates that the U.S. economy is heavily dependent on the flow of goods by sea. If that flow were interrupted, it would not be easy to replace it by other forms of transportation. In fact, the heaviest merchandise goes by sea, and any replacement, if possible, would be extremely costly. For example, the top three maritime trading partners of the United States by weight are Mexico (9.5 percent of total U.S. maritime trade), Venezuela (9.3 percent) and Saudi Arabia (7.4 percent) because oil is the primary commodity coming from these countries. While some of the trade with Mexico could take the continental route (and indeed, since the inception of the North American Free Trade Agreement (NAFTA), in 1994, truck crossings from Canada and Mexico have been increasing at about 5 percent a year), there is no easy replacement in the case of Saudi Arabia or Venezuela (or Germany, the fourth top maritime partner of the United States). Similarly, by value the top three maritime trading partners are Japan (16 percent), China (13.4 percent), and Germany (5.9 percent), which points to the enormous importance of the Pacific sea lanes for the U.S. economy.

Furthermore, commercial interactions link the United States to two strategic regions, Europe and Asia. These regions are analogous in importance to what India and China were for Venice, or the Danube valley for the Ottomans. Any change to these regions, whether caused by wars, domestic strife, or simply the decline of their economies, had an impact on the fortunes of the states linked to them by trade. In the case of the United States, what happens in Europe or parts of Asia will influence U.S. economic well-being. For instance, Japan, the United Kingdom, and Germany are the top three trading partners of the United States by air, followed by other Asian states (South Korea, Taiwan, Singapore, Malaysia). The interruption of air cargo is perhaps less likely than that of sea lanes because it is difficult to threaten and shut down all air routes, but these data stress the importance of East and Southeast Asia for the U.S. economy.[19] Growing political instability in the region, spurred, for instance, by Beijing's increasingly hostile attitude, may affect the ability of Taiwan or Japan to be a key trading partner of the United States. Similarly, the collapse of North Korea (or, alternatively, a war initiated by it) would at a minimum destabilize the economy of South Korea, one of the United States' top ten trading partners. Arguably, the United States could replace one trading partner with another, minimizing the impact of a collapse of one of them. This is what happened to a degree during the Asian financial crisis in 1997, but it was a momentary reduction in trade volume, and by 1999 trade had returned to, or even surpassed, the pre-1997 levels. The collapse of one of these partner economies as a result of a military attack or regional instability, in which case its ability to recover quickly would be limited, would be a greater challenge. In such a case there would be no easy, permanent replacement for, say, Taiwan's or Japan's economy.

The scenarios of instability, threatening both the survival of these U.S. partners and the links by sea (and, to a smaller degree, by air) between them, are several, and it is not my goal to examine them in detail here.[20] The main purpose here is simply to point out that the geopolitical situation—the centers of resources and the trade routes—has a profound influence on the United States. Like its predecessors, the United States cannot ignore it and has to pursue a geostrategy that reflects it. This means maintaining the capability to defend the key routes, mostly sea lanes in the Pacific, and protecting the existence of U.S. trading partners. In this last respect, the United States has an important advantage over the three great powers examined in the book. Venice, for instance, had no ability to influence, or even tap directly, the centers of resources in Asia; it had to make use of a series of intermediaries in Central Asia, the Middle East, and the Indian Ocean. The Ottoman Empire tried to establish direct links with India in the first decades of the sixteenth century by sending an expedition to the Red Sea, India, and the Persian Gulf, but it ultimately failed to establish control over the source of so many luxury goods for Europe's markets. In marked contrast to these cases, the United States is the security provider to most of its trading partners (e.g., Japan, Taiwan, South Korea), as well as the principal guardian of the main sea lanes. In fact, as mentioned above, the United States guarantees the flow of energy, as well as trade, to a potential challenger, China. This gives the United States enormous leverage, something that neither Venice nor the Ottoman Empire, and certainly not Ming China, could claim to possess by the mid-sixteenth century. The United States, has the capability to protect the centers of resources that are important to its economy and, if necessary, to prevent its enemies from accessing them. In other words, the United States is in a position analogous to that of the "Atlantic" powers (Portugal, Spain, and later the Netherlands and Great Britain), which from the sixteenth century on had access to the Asian and American markets, as well as control over the routes to and from them. By controlling these routes, the United States may find it easier to adjust to the shift in the center of resources away from the heart of Eurasia and toward East Asia. It already has some of the strategic tools necessary to maintain its presence and exercise leverage in the new region, unlike Venice or Ming China in Southeast and East Asia in the 1500s.

Nevertheless, the United States cannot ignore the necessity of adjusting its geostrategy. Specifically, it will have to develop two sets of tools to cope with the geopolitical situation that is taking shape. First, it will have to restructure its alliances to reflect the new reality. In particular, if the U.S. objective is to limit China's ability to threaten the Pacific sea lanes and American partners in East Asia, the United States may have to put pressure on China's land borders. Such pressure would force China to direct resources to the defense of its borders rather than to the expansion of control and influence over the South China Sea and further. In other words, the United States would have to implement a strategy similar to that pursued against the Soviet Union when it contained Soviet power through a ring of allies neighboring the Soviet Union (including, from the 1970s on, China). Now alliances with Taiwan and Japan would have to be complemented with strategic partnerships with states on the northern and northeastern borders of China. An opportunity to develop such allies followed the 9/11 attacks on the United States, when American efforts to dislodge the Taliban regime from Afghanistan led to the establishment of military footholds in several Central Asian states (Tajikistan, Kyrgyzstan, Uzbekistan) and to very friendly relations with Russia.[21] The challenge of the next years and decades will be to maintain a U.S. presence in Central Asia and foster a relatively amicable relationship with Russia in order to force China to devote attention and military efforts to its borders. However, U.S. strategy toward both Central Asia and Russia is likely to be driven by short-term concerns (e.g., stability in Afghanistan or the Middle East) rather than by the possibility of a long-term strategic partnership in case of a conflict with China. A strategic partnership with Russia and Central Asia to counterbalance China's rising power in East Asia is therefore difficult to attain, and this difficulty may hamper the United States' ability to contain Chinese ambitions in the Pacific. The alternative is to develop regional, multilateral security arrangements in East and Southeast Asia, which according to most analysts is a desirable goal but one that will not solve the major security problems of the region (such as China's threat to Taiwan).[22] This leaves the United States with a limited array of diplomatic choices that would force China to divert resources away from competing in the Pacific Ocean.

The lack of good diplomatic options available to the United States makes an adjustment in its geostrategy all the more reliant on the ability to retool its military force for the Asian theater. As observed in the historical case studies, new theaters of actions demand new technologies. For example, used to the relatively calm waters and the short distances of the Mediterranean, neither the Ottomans nor the Venetians were prepared to project power to the Indian or the Atlantic Ocean. The technologies at their disposal, the Mediterranean galley, simply did not fit the new geographic theaters and could not compete with the sturdier, oceangoing ships of the Portuguese or Spaniards. Such geographic limitations of technologies continue to be relevant today. For instance, the U.S. military is well prepared to fight a war in Europe but has more limited capabilities in the Asian theater. For one thing, the geographic size of the two theaters is different: Europe east of the Urals is one-quarter the size of Asia. Moreover, U.S. bases in Asia are concentrated in northeastern and southwestern Asia and could be very distant from the area of conflict (e.g., Afghanistan, but also the South China Sea). These distances present a challenge for the U.S. Army, which, having focused on Europe since the Vietnam War, prefers to field heavy forces with long logistic tails, which rely on relatively short distances from U.S. bases. In Asia it may have to deploy far away from bases, increasing the trade-off between speed and weight: the faster a division can be deployed, the less weight it can carry. Air transport is not as readily available as sealift, and it can carry only a limited quantity of supplies and weapons. Consequently, the Asian theater will demand a change in the way the U.S. Army fights wars, which will include innovations in technology (e.g., lighter but equally lethal forces) as well as doctrine (e.g., ground forces may have to fight far away from bases, with limited or no supply lines).[23]

Similar challenges face the air force and the navy. The U.S. Air Force, for instance, has the tools to fight a war in a theater near closely spaced bases or aircraft carriers that are not threatened by the enemy and against a foe that will fight a symmetrical war. Such a situation may not apply to a conflict in Asia, where the potential enemy (e.g., China) is developing "anti-access" capabilities aimed at threatening U.S. bases and aircraft carriers through a combination of ballistic missiles and submarines.[24] Moreover, potential U.S. challengers do not plan to match the technological superiority of American airplanes but simply to develop relatively low-cost solutions (e.g., missiles or unmanned aircraft) to strike U.S. forces from a distance.[25]

An analogous problem faces the U.S. Navy, which during the cold war planned to fight a war against a comparable navy and to play a logistical-support role to ground forces engaged on land in Europe. In Asia the strategic situation is likely to be profoundly different. In the next decades, in part because of fiscal and technological constraints, China will not develop naval capabilities comparable to those of the United States. To achieve its strategic objectives (e.g., control over

Taiwan), China can simply threaten the U.S. naval presence in the Taiwan Strait and the South China Sea. Recent acquisitions of *Sovremenny*-class destroyers and *Kilo*-class submarines seems to indicate precisely such an attempt to challenge local U.S. maritime supremacy without matching the global reach of the U.S. Navy.[26] In the case of a conflict with China the U.S. Navy will have to fight in coastal waters defended by submarines as well as land-based missiles and planes. In such circumstances the vulnerability of most naval platforms, and of the carrier in particular, will be so marked as to put in doubt their usefulness. Technological innovations (e.g., increased antisubmarine capabilities or antimissile protection) or outright new technological and doctrinal approaches (e.g., replacement of the carrier with smaller, faster ships serving a missile platforms or preemptive strikes against land targets) will be necessary to prepare U.S. forces for the Asian theater.

Finally, the United States must keep in mind that its ability to project power abroad is conditional on the stability of its land borders, or, broadly speaking, on its territorial security. Surrounded by oceans and weak states, the United States has benefited from decades of territorial security. As a result of this stability, it could project power far from its borders, unlike continental powers such as Russia and Germany, whose ability to devote resources to noncontiguous regions was severely hampered. The presence of a potential source of trouble on the southern frontier (e.g., Mexico could have become a strategic front during World War I had the German diplomatic advances been successful) has never provoked a sustained fear for the stability of U.S. land borders. The border with Mexico is, however, an important variable because its situation could affect the United States' ability to be engaged in the world. If, for instance, Mexico collapses as a viable state, resulting in an enormous flood of refugees, the United States might have to redirect some of its military capabilities to the protection of its territory. Although this is, admittedly, a distant and unlikely scenario, it should not lead to complacency. The territorial security the United States has enjoyed over the past century is not a historical constant, and as 9/11 demonstrated, it has to be protected from a variety of threats.

The list of diplomatic and technological challenges associated with geostrategic adjustments is long, and it conveys the difficulty of adapting to geopolitical shifts. The United States, like preceding great powers, cannot rely exclusively on its power to sustain its position of supremacy because power is only a tool, and as such it can be misused. In fact, states often pursue a foreign policy that does not reflect the underlying geopolitics and use power in areas that bring no strategic gains, leading to their weakening and even demise. This is why great powers, now as in the sixteenth century, are precarious entities whose existences, as St. Augustine observed, "may be compared to glass in its fragile splendor."[27] A slight change in geostrategy redirecting the attention and energies of the state away from strategic regions can lead to decline.

To maintain a connection between geopolitics and geostrategy, in the next years and decades the United States will have to retool its foreign policy to reflect the shift of power toward East Asia. Failure to do so, as history teaches us, leads to decline and ultimately to strategic irrelevance, as in the cases of Venice and the Ottoman Empire, or collapse, as in the case of Ming China.

Introduction

1. Obviously, the questions how to get power and what to do with it are interrelated. Often the answer to the latter carries an answer to the former. For instance, where a state projects its power can determine its relative strength because the extension of control over regions that are strategically irrelevant leads to an expenditure of power with no tangible returns. However, these are analytically distinct questions that can be examined separately.

2. Montesquieu, *Considerations on the Causes,* 169.

3. Moreover, these historical cases are rarely studied from the perspective of international relations and U.S. foreign policy. But as I argue, they offer important lessons on the importance of a geostrategy that reflects geopolitics. They are an interesting and important addition to the case studies used in international relations theory.

4. The literature on this subject is vast. See, e.g., Tilly, *Formation of Nation States;* idem, *Coercion, Capital, and European States;* Małowist, *Europa i jej ekspansja;* and Gilpin, *War and Change,* esp. chap. 3. For a more recent treatment of the idea of the nation-state and its effectiveness, see Fukuyama, *State-Building.*

5. C. Vann Woodward, for instance, argues that the "free security" enjoyed by the United States in the nineteenth century contributed to the development of a belief in the "innocence" of the country. Power was not needed by the United States; it was something that only the "wicked world" used. See Woodward, "Age of Reinterpretation." See also Wolfers, *Discord and Collaboration,* 233–51; and Albrecht-Carrié, *Diplomatic History of Europe,* xiii–xvi.

One • The Premature Death of Geography

1. Tocqueville observed, "It is unbelievable how many systems of moral and politics have been successively found, forgotten, rediscovered, forgotten again, to reappear a little later, always charming and surprising the world as if they were new, and bearing witness, not to the fecundity of the human spirit, but to the ignorance of men." Quoted in Yost, "Political Philosophy," 265. Geography seems to be one of those forgotten systems.

2. For instance, in 1989 the University of Paris established a program in geopolitical

studies, offering even a PhD in geopolitics. A similar renaissance of geopolitics occurred in other European countries. In Italy in 1991 a review of geopolitics, *Limes,* began to be published.

3. The most common criticism of the idea of a renaissance of geography is that "geography no longer matters" because of the power of new technologies. In other words, geography has disappeared from theory because it has lost its influence over political reality. Part of this criticism cannot be satisfied because of the different set of assumptions concerning the nature of states and of their political interactions. Such criticism is grounded in the view that power and its application can be detached, however gradually, from its spatial dimension. In my view, states are now and always will be territorial animals regardless of the cybercommunication technologies, economic "globalization," or global military capabilities. They act within an environment that is exogenous, that needs to be accepted as given, and that cannot be rendered completely irrelevant. Political actions in their economic or military forms are tied to geography; power can be exercised only within a spatial dimension.

Part of the criticism is based on a hypothesis that there is an inverse relationship between the importance of geography and technology (or more broadly power): when technology increases, geography decreases. I argue instead that technological advances bring to light different geographic constraints by altering patterns of movement and the location of strategic resources. As I examine in greater detail later, geography, conceived as the geological distribution of land and seas, of mountain ranges and rivers, is immutable, but its political importance changes on the basis of technological advances. Therefore, technology constantly shapes geography, making it neither constant nor increasingly irrelevant for international relations.

4. The study of geography and politics has a long history, dating back to ancient philosophers and historian such as Aristotle and Pliny. However, because my goal is not to describe the intellectual history of geography and politics but to determine the reasons behind the modern unpopularity of geopolitics, I limit my study to the more recent literature on the subject. See Fairgrieve, *Geography and World Power;* Moodie, *Geography behind Politics;* and East, *Geography behind History.* Also, the most well known geopolitical writer, Sir Halford Mackinder, was a geographer by training. On the role of geographers in the origins of geopolitics see also Clokie, "Geopolitics," 493.

5. On the influence of climate on political decisions and performance see also Sprout and Sprout, *Foundations of International Politics,* 340–64; and Konigsberg, "Climate and Society." For a quick overview of the "ancients" see also Cohen, *Geography and Politics,* 24–35; and Gottmann, "Background of Geopolitics."

6. On the configuration of seas see Mahan, *Influence of Sea Power upon History.* On the length and direction of navigable rivers, see, e.g., McNeill, "Eccentricity of Wheels." And on the disposition of continents see Diamond, *Guns, Germs, and Steel.*

7. Clokie, "Geopolitics," 493.

8. Dugan, "Mackinder and His Critics Reconsidered." From the beginning of his career Mackinder was interested in making geography "a discipline instead of a mere body of information" and consequently in making it more relevant to "the practical requirements of the statesman and the merchant." See Mackinder, "Scope and Methods of Geography."

9. Mackinder, "Geographical Pivot of History," reprinted in Mackinder, *Democratic Ideals and Reality*, 176.

10. Mackinder, *Democratic Ideals and Reality*, 2.

11. "My contention is that from a geographical point of view they are likely to rotate round the pivot state, which is always likely to be great, but with limited mobility as compared with the surrounding marginal and insular powers." Mackinder, "Geographical Pivot of History," reprinted in ibid., 192.

12. "The oversetting of the balance of power in favor of the pivot state, resulting in its expansion over the marginal lands of Euro-Asia, would permit the use of vast continental resources for fleet-building and the empire of the world would be in sight. This might happen if Germany were to ally herself with Russia. The threat of such an event should, therefore, throw France into alliance with the over-sea powers, and France, Italy, Egypt, India, and Korea would become so many bridge heads where the outside navies would support armies to compel the pivot allies to deploy land forces and prevent them from concentrating their whole strength on fleets." Ibid., 191–92.

13. Mackinder, *Democratic Ideals and Reality*, 150.

14. In his many revisions of this concept Mackinder considered the role played by technological advances in military and communication means that had a "shrinking" effect on space, making the Heartland increasingly larger. In particular, railways had an impact on the land power equivalent to that of steam on the sea power, increasing the speed and mobility of goods and troops. But the Heartland continued to be a distinct geographic area that no technological advance could make irrelevant. As Mackinder points out, the pivot region of the world's politics is that "vast area of Euro-Asia which is inaccessible to ships, but in antiquity lay open to the horse-riding nomads, and is today about to be covered with a network of railways." Mackinder, "Geographic Pivot of History," reprinted in Mackinder, *Democratic Ideals and Reality*, 191. See also Cohen, *Geography and Politics*, 42–44; and Sloan, "Sir Halford Mackinder."

15. Raymond Aron offers a different interpretation of Mackinder's thesis. In Aron's view, Mackinder speaks of geographic causality and determinism without supporting his views. In fact, Aron argues, Mackinder's principal point is that human beings adapt to geographic facts and conquer them through technological innovations. Hence, there are two perspectives on the world that depend on the way men adapted to geography: those of the "seamen" and the "landmen." Aron argues that because geography does not determine the success of one over the other, Mackinder believed in the determining role of human actions rather than in that of geography. In other words, while Mackinder is outwardly a natural scientist, at heart he is a social scientist. See Aron, *Peace and War*, 197–98. While I share Aron's observation on the importance of the two "perspectives" on the world and the shaping role of human inventiveness, I disagree with his conclusion that Mackinder favored human actions over geography. In fact, Mackinder's fundamental world-view was heavily deterministic and rendered inflexible by his stress on the role played by the immutable and unchangeable geographic factors. As a result, as Harold and Margaret Sprout observed, Mackinder "persisted to his dying day in the attempt to fit these [technological advances] and other changes into an increasingly obsolescent frame of reference in which the most strategic variables were location, space, distance, and geographical configura-

tions." Sprout and Sprout, "Geography and International Politics," 152. See also Howard, "Influence of Geopolitics."

16. This branch started in Germany with writers like Friedrich Ratzel (1846–1911) and his Swedish follower Rudolf Kjellén (1864–1922), who first coined the term *geopolitics*, and continued with Karl Haushofer (1869–1946). See also Herwig, "*Geopolitik*"; Kiss, "Political Geography into Geopolitics"; Whittlesey, *German Strategy of World Conquest*, 70–78; and idem, "Haushofer." Interestingly, Whittlesey's chapter was dropped in later editions of *Makers of Modern Strategy*, in which there is no mention of geopolitics. Edward Mead Earle, editor of that volume, also expressed serious reservations regarding the introduction of geopolitics in the United States by Nicholas Spykman in the 1940s. See Earle, "Power Politics and American World Policy."

17. Vagts, "Geography in War and Geopolitics," 84.

18. On the use of geography as a justification for an ideology and as a tool of political propaganda see Speier, "Magic Geography."

19. Kristof, "Origins and Evolution of Geopolitics," 20.

20. Kiss, "Political Geography into Geopolitics," 645.

21. Clokie, "Geopolitics," 497. The passage from geography to policy began at the end of the eighteenth century, when political geography "passed . . . into the realm of practice in certain countries and into the realm of philosophy in others." On the one hand, Ratzel introduced concepts of space and location that bordered on the metaphysical. On the other hand, Vidal de la Blanche emphasized that history was moved by social forces. See Gottmann, "Background of Geopolitics," 200.

22. Gottmann, "Background of Geopolitics," 200.

23. Bowman, *Geography in Relation to the Social Sciences*, 212.

24. See also Dorpalean, *World of General Haushofer*, 38–42; Kruszewski, "International Affairs"; Gyorgy, "Application of German Geopolitics"; and Strausz-Hupé, *Geopolitics*. For a detailed examination of the intellectual roots and ideological products of geopolitics see Wolff-Poweska, *Doktryna geopolityki w Niemczech*. And for a concise criticism of geopolitics as a "single-factor explanation" see Morgenthau, *Politics among Nations*, 116–18.

25. Bowman, "Geography vs. Geopolitics," 657.

26. Already in the aftermath of World War I a breach between the deterministic, natural scientific approach and the more sociohistorical approach to geopolitics had become visible. In response to the German approach, the French school of geopolitics, with writers like Vidal de la Blanche, argued that the state was "a product of history, a human rather than a natural creation." But it was only after World War II that the balance tilted toward the less deterministic approach. See Gottmann, "Background of Geopolitics," 202.

27. See Hoffman, "American Social Science," 43–44. For a brief synopsis of early American geopolitical writers see Kristof, "Origins and Evolution of Geopolitics," 37–42. Clokie calls Spykman "a 'political scientist' who resorts to geography to illuminate his subject." Clokie, "Geopolitics," 497.

28. Quotation from Morgenthau, *Politics among Nations*, 106. In fact, Morgenthau considers geopolitics one of the three single-factor explanations of national power (the other two are nationalism and militarism) that, by "attributing to a single factor an over-

riding importance, to the detriment of all the others," lead to profoundly mistaken assessment of the balance of power. Ibid., 153–54.

29. Ibid., 108. See also Stephen Jones, "Power Inventory and National Strategy," 447.

30. Spykman, "Geography and Foreign Policy, I," 32. Morgenthau is somewhat contradictory on this point. While arguing that Russia defeated Napoleon and Hitler thanks to its vast spaces, he asserts later on that large territories do not necessarily mean a more powerful state. Morgenthau, *Politics among Nations*, 151–52.

31. Morgenthau, *Politics among Nations*, 108.

32. Ibid., 106, 107–8.

33. Spykman, "Geography and Foreign Policy, I," 41; idem, *America's Strategy in World Politics*, 446–47. For instance, location on the shore of an open sea or near important routes of communications determines the fate of a state. When the trade routes change their pattern, the state located astride them is condemned to loose its importance. See Gilpin, *War and Change*, 112–13.

34. Spykman, "Geography and Foreign Policy, II," 224.

35. Harold Sprout, "Geopolitical Hypotheses in Technological Perspective," 192.

36. See also Spykman, *America's Strategy in World Politics*, 447.

37. For "land mysticism" see Deudney, *Whole Earth Security*, 9. For contradictions on the importance of territorial size, see also n. 30 above.

38. Morgenthau, *In Defense of the National Interest*, 10.

39. Ibid., 175–76; Morgenthau, *Politics among Nations*, 81.

40. Spykman, *Geography of the Peace*, 43.

41. See, e.g., Morgenthau, *Politics among Nations*, 145–49. Geographical features make the calculation of power more difficult and uncertain.

42. In such a model the only variable is power: in Wolfers's words, "Expansion of power at the expense of others will not take place if there is enough counterpower to deter or to stop states from undertaking it." Wolfers, "Pole of Power and the Pole of Indifference."

43. Spykman, *America's Strategy in World Politics*, 165.

44. For a model of the loss-of-strength gradient, see Boulding, *Conflict and Defense*, 245–47.

45. Spykman, *America's Strategy in World Politics*, 165.

46. See also Aron, *Peace and War*, 182–83.

47. Sprout and Sprout, *Toward a Politics of the Planet Earth*, 24.

48. Wohlstetter, "Illusions of Distance."

49. Sprout and Sprout, *Foundations of International Politics*, 339.

50. Singer, "Geography of Conflict," 1.

51. Waltz, *Theory of International Politics*, 9.

52. The desire to abandon geography is owing in part to the motivation behind a theory. As Waltz says, the desire behind any theory is the desire to control, not only to predict. Ibid., 6. Because it is impossible to control the disposition of continents or the patterns of climate, it is easier to abstract from them and consider only those aspects that can be controlled. States and men can only adjust themselves to geography; rarely can they control it (an exception might be, for instance, the diversion of water through dams).

53. Furthermore, in addition to the trend of making geopolitics social scientific, there is a tendency to generalize the American political experience characterized by distant threats and easily reachable strategic objectives. The United States has been blessed with a geopolitical position that allowed it to remain separate from the troubles of Eurasia without depriving it of the option of global involvement. Coupled with a superpower's global reach, this geopolitical blessing has engendered an inclination to discount the influence of geography on the fate and interests of a state. This American perspective on geography has heavily influenced the theory of international relations.

54. For example, there has been a revival of geography in economics. While some studies stress the economic impact of environmental characteristics (e.g., climate or access to seas or rivers), most treat geography in relational terms: the value of a place is measured not according to its own features (natural resources or climate) but according to its location in relation to other markets and centers of production. The most interesting insights on the role of such a concept of geography come from a long tradition that began with Alfred Marshall and is currently promoted by Paul Krugman. These scholars ask why production locates in a specific place, with the resulting concentration of industry, like the Detroit auto industry or Silicon Valley. There are no apparent advantages associated with those locations for the type of economic activity undertaken. While randomness plays a great role in the location of the first firm, Marshall and Krugman give three reasons for the further concentration of production. First, a cluster of firms is more able to support specialized suppliers; second, it allows for labor-market pooling, which is beneficial to both employers and workers; third, it results in a highly beneficial knowledge spillover that lowers the entry costs for new firms. The result is a high concentration of industry in the same location. Such an approach to geography points to the idea of geographic changes. At first this appears to be a contradiction in terms given that the very nature of geography implies immutability. Yet, because geography is seen in relational terms, locations change in value because of changes in markets, resource utilization, and transfer technologies. Hence, a modification in the path of a trade route can dramatically change the economic viability of a city or a state; a seaport can have the best natural bay, but if its consumption or production markets change and move far away, its geographic advantages are of no use. Because economic development is geographically differentiated and continues to change its geographic loci, the importance of places fluctuates according to the rise and decline of centers of economic wealth. See Krugman, "Increasing Returns and Economic Geography"; idem, *Geography and Trade;* Venables, "Assessment"; and "Knowing Your Place." For a more deterministic perspective on the role of geography in economics see Sachs, "Nature, Nurture, and Growth." For a classical economic treatment of geography see also Hirschman, *Strategy of Economic Development,* 183, 185–99; and Hoover, *Location of Economic Activity,* 10.

55. For a good overview of the debate on neorealism see Keohane, *Neorealism and Its Critics.*

56. "Structurally we can describe and understand the pressures states are subject to. We cannot predict how they will react to the pressures without knowledge of their internal dispositions. . . . Systems theories, whether political or economic, are theories that explain how the organization of a realm acts as a constraining and disposing force on the inter-

acting units within it. Such theories tell us about the forces the units are subject to." Waltz, *Theory of International Politics*, 71–72.

57. See, e.g., Brzezinski, *Game Plan;* idem, *Grand Chessboard;* Chase, Hill, and Kennedy, *Pivotal States;* and Kemp and Harkavy, *Strategic Geography.*

58. See Rose, "Neoclassical Realism." See also Zakaria, *From Wealth to Power,* 21–31. For an early statement of the "security" hypothesis see Herz, "Idealist Internationalism." For a more recent one see the classic article by Jervis, "Cooperation under the Security Dilemma"; and idem, "From Balance to Concert."

59. Walt, *Origins of Alliances,* 23.

60. Ibid., 277; see also ibid., 23–24.

61. Distance is the "ultimate geographic variable" (Sprout and Sprout, *Foundations of International Politics,* 339), but it oversimplifies geographic realities. See below on offense-defense.

62. See also a critique of Walt's theory by Robert Kaufman, "To Balance or to Bandwagon?"

63. Jervis, "Cooperation under the Security Dilemma," 173–74. This observation is similar to the one made by Walt, the difference being the dependent variable: Walt is concerned with state behavior (foreign policy), while Jervis is more interested in the system.

64. Van Evera, "Offense, Defense," 19. For a conceptual clarification of the offense-defense balance see Glauser and Kaufmann, "What Is the Offense-Defense Balance?" For a critique and an overview see Levy, "Offensive/Defensive Balance."

65. Jervis, "Cooperation under the Security Dilemma," 195.

66. Hopf, "Polarity," 476.

Two • *Geography, Geopolitics, and Geostrategy*

1. According to Saul Cohen, "Geography can be defined as the *science of area differentiation.* Its essence is to observe, inventorize, map, classify, analyze, and interpret patterns of earth-man relationships over different parts of the earth's surface.... Geographers seek out subdivisions within the physical environment (climate, soils, vegetation, and landform) and subdivisions in the cultural or man-made environment." Cohen, *Geography and Politics,* 3.

2. Disasters are defined as "abrupt, major, negative shocks which reduce the aggregate assets or income of a given population." Eric Jones, "Disasters and Economic Differentiation," 677. A simple analysis of environmental disasters is by itself a powerful explanatory variable of, for instance, the growth differential between Asia and Europe. "Europe is and was a safer piece of real estate than Asia." Ibid., 681.

3. The continental expansion of the United States is an excellent example of a geographic change to a state's setting. As Saul Cohen observes, the "geographical setting of a state and its global outlook are interrelated. As settings change, so must these outlooks." Cohen, *Geography and Politics,* 93–94. Hence, because of the different historical periods, writers like Arnold Guyot, Alfred Mahan, and Nicholas Spykman forecasted different geostrategic inclinations of the United States.

4. Instead of *geopolitics* some use the term *strategic geography.* See, e.g., Gooch, "Weary Titan," 282–83; and Kemp and Harkavy, *Strategic Geography.*

5. By *strategic importance* I mean the impact the control, or lack of control, of such locations would have on a state. The strategic importance of locations is dependent on the presence of key resources (e.g., oil or industrial production) or on their controlling position in relation to trade routes (e.g., straits, islands, valleys, etc.). It changes following shifts in routes and in the value of resources.

6. Because geopolitical changes occur over decades and centuries, it is also difficult to pinpoint their causes. "Big processes call for big causes" is the "golden rule of historical analysis." Landes, "What Room for Accident in History?" 653. See also Christian, "Case for 'Big History.'" As a result parsimony is sacrificed because the complexity of a big process requires the study of multiple causes, on different levels of analysis and time. In the study of geopolitics, therefore, monistic explanations are most likely to be wrong. This is not to say that big trends can be the result of small decisions, taken without any consideration or awareness of the long-term consequences they engender. For instance, Columbus's decision to sail westward from Europe was a small decision that had enormous and long-term consequences. But alone it was not sufficient to generate a shift in power; it had to be followed by decades of exploration, conquest, and imperial rule. Only the sum of all these actions resulted in the sixteenth-century geopolitical change that undermined the importance of the Mediterranean. As David Landes writes, "Historians tend to be suspicious of simplicity; they see it less in events and developments than in the eye of the beholder. Large processes of historical change are not likely to hinge on single causes; many pieces have to come together. Hence, the presence or absence of particular features in other places should not lead us to expect or preclude parallel and simultaneous processes of development" (653).

7. Harold and Margaret Sprout call this hypothesis "environmental possibilism" and place it in the middle of a spectrum ranging from "environmental determinism" (milieu absolutely determines the outcome) to "cognitive behavioralism" (men react to the milieu as they perceive it). See Sprout and Sprout, *Man-Milieu Relationship Hypotheses*, 40.

8. For the classic statement of the role of perception in shaping the evidence see Jervis, "Hypotheses on Misperception."

9. Spykman, "Geography and Foreign Policy," I, 43. See also Henrikson, "Map as an 'Idea'"; and idem, "Geographical 'Mental Maps.'"

10. Quincy Wright writes that "the apparent permanence of geographical conditions is illusory. The human significance of geography depends on the conditions of society, population, and technology. Civilized man uses his environment to serve ends which come from other sources. He must know the properties of his position, terrain, climate, and resources, as the chemist must know the properties of the elements. These properties can be put to many uses, can serve many ends, and can support many types of international relations. Geography cannot develop concepts and conceptual systems applicable beyond a limited time and area in which a given state of the arts, of population, and of society can be assumed." Quincy Wright, *Study of International Relations*, 348.

11. "Like the tectonic forces that move continents around on the surface of the earth, historical tectonics lie beyond our normal range of perception. No single nation or individual sets them in motion; they result, rather, from the interaction of events, conditions, policies, beliefs, and even accidents. They operate over long periods of time, and across the

boundaries we use to define place. Once set in motion, they are not easily reversible."
Gaddis, *United States and the End of the Cold War*, 156.

12. Specific configurations of routes and locations of resources often define historical periods, such as the sixteenth-century "Vasco da Gama" era or the rise of the Atlantic world that challenged the primacy of the Mediterranean.

13. Often the definition of *geopolitics* is left very vague. Saul Cohen, for instance, defines it as "the relation of international political power to the geographical setting." Cohen, *Geography and Politics*, 24. In large measure, the elusiveness of such definitions is meant to circumvent the danger of linking the constantly changing international relations or the foreign policies of states to an immutable geographic reality.

14. Routes also are vital for the internal economic and political cohesiveness of a state. Internal routes are important in uniting the different regions of the state and connecting them with the center. The Roman Empire, with its imperial routes used by legions and merchants, is perhaps the clearest example of the political and military importance of internal routes. The decay of routes is often a sign of a weakening central authority. See also Spykman and Rollins, "Geographic Objectives in Foreign Policy, II."

15. U.S. Department of Transportation, Bureau of Transportation Statistics, *G-7 Countries*, 29. It is also significant that the is one of the greatest maritime commercial powers: in 1997 21 percent of the global waterborne trade was American. U.S. Department of Transportation, Bureau of Transportation Statistics, Maritime Administration, U.S. Coast Guard, *Maritime Trade and Transportation, 1999*, xi.

16. Even in the "missile age," when nuclear weapons can be delivered anywhere in the world, the capacity to project conventional forces is indispensable to convey the commitment to the region in question. As Michael Howard observed, "The more remote a crisis or a country from the territory of a nuclear power, the more necessary it will be for that power to deploy conventional forces if it wishes to demonstrate the intensity of its interest in that area." Howard, "Relevance of Traditional Strategy," 264. For an interesting analysis of why bases on sea lanes are important in the nuclear age see Blechman and Weinland, "Why Coaling Stations Are Necessary"; and Wohlstetter, "Illusions of Distance," 248–49. At the turn of the twentieth century Lord Fisher identified five key bases that controlled the world: Singapore, the Cape, Alexandria, Gibraltar, and Dover. See Friedberg, *Weary Titan*, 200.

17. Van Creveld, *Technology and War*, 181.

18. Paradoxically, the more advanced the army, the more logistical support it needs, making it more dependent on lines of communication. In other words, technological advances in the military field are increasing the importance of routes through which an army is supplied.

19. Gilpin, *War and Change*, 112.

20. At that time Russia had three fleets: the Pacific Fleet, effectively destroyed or incapacitated by the Japanese; the Black Sea Fleet, locked in the Black Sea because of the 1841 Straits Convention, forbidding the passage of warships through the Bosphorus in time of war; and the Baltic Fleet. See Seton-Watson, *Russian Empire*, 591.

21. For a lively account of the voyage of the Baltic Fleet see also Hough, *Fleet That Had to Die*. For the role played by the Hamburg-America Line see Cecil, "Coal for the Fleet"; Walder, *Short Victorious War*, 185–86.

22. The kaiser suggested that Germany's help to Russia in the form of coal supplies was conditional on some sort of defensive alliance between Moscow and Berlin against a potential attack by Great Britain. White, *Diplomacy of the Russo-Japanese War*, 183.

23. This conclusion was reached immediately after the war by a French military analyst who argued that Russia had missed its opportunity to establish maritime bases and link its main fleets. Daveluy, *La lutte*, 1–2, my translation.

24. Mahan, *Influence of Sea Power upon History*, 25.

25. See, e.g., Rhodes, "' . . . From the Sea' and Back Again."

26. On the idea of "natural routes" see East, *Geography behind History*, 56–73.

27. Before the Panama Canal was finished, Mahan wrote that such a canal would change the Caribbean "from a terminus, and a place of local traffic, . . . as it is now, into one of the great highways of the world." Mahan, *Influence of Sea Power upon History*, 33. This would also alter the geopolitical position of the United States, making it similar to that of England in relation to the Channel or of the Mediterranean countries in relation to the Suez Canal.

28. Technology does not "conquer" geography but merely alters the pattern of routes and resources. Technological evolutions (or revolutions) alter the way geography is "used" for the production of energy and for the transport of goods and information, but they do not erase geographic constraints altogether. They only alter the importance attached to locations. The belief that technology makes geography increasingly irrelevant appears to be stronger at the onset of a new technological era, when the invention of a new means of communication or of a new form of energy production is heralded as a further, if not final, conquest of geography. As time passes, such a belief loses potency for a variety of reasons: the technology spreads and erases the advantages of one state, leading to a renewed status quo; new technologies are devised to serve as countermeasures; and so on. The excitement about a new technology, however, arouses anew the belief in a further depreciation of and freedom from geography.

29. Often new technologies do not increase but limit the range of possible routes. For instance, following the political instability in the region of the Suez Canal in the 1960s and 1970s, Western powers, with Japan at the technological forefront, introduced the very large crude carriers (VLCs), or supertankers. The VLCs made long-haul transport of oil more cost effective, and the oil tankers could avoid the shorter but unstable route through the Suez Canal by circumnavigating Africa. Once the political situation between Egypt and Israel stabilized, however, many supertankers turned out to be too large to pass through the canal and were forced to continue to navigate the longer and thus more expensive route around Africa. For instance, the Ras Tanura–Rotterdam route via the Suez Canal (12,698 miles) is 76 percent shorter than the Cape route (22,338 miles); the Yokohama-London route through the Canal is 31 percent shorter (22,042 miles versus 28,854 miles). See "How to Shorten Tanker Voyages"; see also Yergin, *The Prize*, 496–97.

30. Margaret Sprout, "Mahan," 424. On the importance of railways see also Earle, "Adam Smith, Alexander Hamilton, Friedrich List."

31. Paul Kennedy, *Rise of the Anglo-German Antagonism*, 44. The same could be said for the relations between Great Britain and Russia. As John Gooch observes, for the British, "technology . . . dictated the end to any residual thoughts of repeating the Crimean War;

the completion of the Trans-Caspian railway line placed the Russian supply route to Afghanistan beyond the reach of any British expedition to the Black Sea." Gooch, "Weary Titan," 282.

32. Here I do not examine routes that are or have been important for the internal unity of the state. Such routes, for example, those of the Roman and Inca empires, connect the periphery of the state with the center in order to facilitate the centralization of political control. Roads unify people and help the administration of the state by "integrating the territory," and from this perspective railroads are considered to be a crucial factor in the integration of vast continental states. As Spykman observes, "Only effective integration of territory permits successful resistance to centrifugal tendencies at the periphery and pressures on far-flung frontiers." The collapse of the empires has usually been accompanied by the decay of these roads and the restoration of geographic barriers to communication. Spykman, "Geography and Foreign Policy, II," 226; see also idem, "Geography and Foreign Policy, I," 36–37.

33. Russia wants the bulk of oil to go through areas under its control, ending on the Black Sea (in ports such as Novorossiysk) or in the domestic pipeline system, from which it can be dispensed to Western Europe. Turkey prefers to see a pipeline linking Baku in Azerbaijan with its Mediterranean port of Ceyhan, while other states are proposing alternate routes through Iran, Kazakhstan, and, further east, Afghanistan and India.

34. See Aron, *Peace and War,* 54; and Morgenthau, *Politics among Nations,* 109–14.

35. I do not include other variables commonly associated with material power, namely, military strength. Although it is an important component of a state's power, military strength by itself does not define geopolitics. Military power in fact is important insofar as it protects or threatens a center of economic or natural resources and the lines of communication. If, for instance, a state that has a sizeable military force is neither a center of resources (or near such a center) nor astride vital routes, it is strategically important only insofar as it can project its military power toward resources and routes. That is, to be geopolitically relevant, a militarily powerful state must present a threat to resource-rich territories and trade routes. In addition, broadly speaking, military power coincides with resources. Especially in the modern era, since the seventeenth century, the coincidence between wealth and power has become stronger. The growing sophistication of military technology has required a correspondingly wealthier state; only the wealthiest states could afford to build and field large armies with the latest technologies. So economic and natural resources are a good proxy for military capability and potential. See Gilpin, *War and Change,* 124–25.

36. Machiavelli, *Discourses,* bk. I, chap. I, p. 108.

37. Morgenthau, *Politics among Nations,* 109. For Morgenthau there are two types of resources: natural resources (foodstuffs and raw materials) and industrial capacity. They are a "relatively stable" component of power. See ibid., 109–14.

38. Rosecrance, *Rise of the Trading State;* Nye, "Changing Nature of World Power"; Brooks, "Globalization of Production."

39. Thompson and Zuk, "World Power."

40. Gilpin, *War and Change,* 138.

41. The question whether conquest pays is important for predicting the response of states. If the conquest of industrial resources does not bring a gain to the offensive power,

the international system might be more stable because states will choose not to use force to obtain the resources they need. Conversely, if it continues to be profitable to conquer resource-rich territories, states will use force to control and exploit natural and economic resources. See Liberman, "Spoils of Conquest"; and idem, *Does Conquest Pay?*

42. Knorr, *Power of Nations,* 14, 80.

43. Kennan argued that there were five vital centers: the United States, Great Britain, Germany and Central Europe, the Soviet Union, and Japan. At the time of this writing, only one center was under Soviet control, and Kennan called for the defense of the other four. His view was subsequently abandoned in NSC-68, which document expanded the geopolitical view by introducing the psychological dimension and arguing that threats to U.S. credibility were as dangerous as those to the centers of power. Gaddis, *Strategies of Containment,* 30–31, 90–92. See also Gerace, "Between Mackinder and Spykman."

44. As Quincy Wright observes, "One environment is more suitable for power development than is another. Furthermore, power may shift because of the exhaustion of resource or perhaps because of climatic changes. Major factors in these shifts, however, have been changes in the arts, in the economic and military value of resources, in the growth of population, and in skills in utilizing resources and in organizing political and military power. Geography can assist in interpreting the historical record, but the nongeographical factors are so important that it is unlikely that geography will ever be able to provide principles or evidence for predicting where the major centers of power will be in the future." Quincy Wright, *Study of International Relations,* 348.

45. Jones, Frost, and White, *Coming Full Circle,* 80.

46. On the importance of water as a strategic resource see Allan, "Virtual Water," 545. For an illustration of water as a weapon in relations between Iraq and Kuwait see Tétreault, "Autonomy, Necessity, and the Small State," 583.

47. The key shift occurred during World War I, when the navies of the belligerent countries, especially Great Britain and Germany, noticed the incredible advantages of oil-fueled ships, which, in comparison with their steam-engine predecessors, had a double radius of action and were considerably easier and quicker to refuel (it took twenty-four to thirty-six hours to coal a ship). Furthermore, the use of oil allowed the development of three new lethal weapons: the tank, the airplane, and the submarine. See also Sédillot, *Histoire du pétrole,* 153–55; and Paul Kennedy, *Rise and Fall of British Naval Mastery,* 205–37.

48. The supply of oil increased from 3.9 million tons in 1880 (60 percent from the United States, mostly Pennsylvania) to 51 million tons in 1913, with an increase of almost 10 million tons between 1910 and 1913. Sédillot, *Histoire du pétrole,* 114–15.

49. In the United States oil constituted 15.3 percent of all energy sources in 1920 (with coal at 78.4 percent) and 37.6 percent in 1951 (with coal at 35.8 percent). Tétreault, *Revolution in the World Petroleum Market,* 20.

50. Odell, *Oil and World Power,* 188.

51. As the British Admiral John Fisher, the "godfather of oil," observed in 1913, when the Royal Navy was starting to switch to oil combustion engines, "Oil don't grow England." Hence, oil had to be found elsewhere and brought back to England. See Yergin, *The Prize,* 157. A similar change in the strategic map can be observed in present-day China, which has recently become a net oil importer. Given that the only source of oil that can satisfy Chinese

needs is in the Middle East and, to a smaller degree, in the Caspian Sea region, the strategic map faced by Beijing is different from that of ten years ago.

52. Diamond, *Guns, Germs, and Steel.*

53. See, e.g., Frank, *ReOrient;* and Pomeranz, *Great Divergence.*

54. This is not to say that natural resources automatically lead to economic strength. In the American case, in fact, the exploitation of the geological endowment "was itself largely an outgrowth of American industrial success." See Gavin Wright, "Origins of American Industrial Success," 666; and Nelson and Wright, "Rise and Fall of American Technological Leadership."

55. See Hausmann, "Prisoners of Geography"; Gallup, Sachs, and Mellinger, "Geography and Economic Development"; Gallup and Sachs, "Location, Location"; and Diamond, *Guns, Germs, and Steel.*

56. For a critique see Bergère, "On the Historical Origins of Chinese Underdevelopment"; and Elvin, "Why China Failed."

57. Friedberg, *In the Shadow of the Garrison State;* Acemoglu, Johnson, and Robinson, "Reversal of Fortune."

58. Gilpin, *War and Change;* Paul Kennedy, *Rise and Fall of the Great Powers.*

59. Jones, Frost, and White, *Coming Full Circle,* ix.

60. Another plausible impediment to the projection of power to distant geopolitically vital areas is the domestic structure of the state. States that lack domestic authority and base their existence on force require a military force whose main goal is the maintenance of domestic order, not the extension of state influence abroad. The military organization and capability, as well as the diversion of economic resources, that such a domestic system demands is a considerable structural hindrance to power-projection capabilities.

61. A more powerful critique of the importance of borders is the argument that some of the greatest threats to the territorial security of a state can come from nonadjacent countries, especially those with missile capabilities. For instance, the development of missile technology by a threatening but distant power, such as China or North Korea, presents serious security challenges to U.S. territory that no longer can be countered on its borders. In other words, even when borders are stable and secure the territorial security of the state can be threatened. The ballistic-missile threat is similar to naval or air threats, which can disrupt commerce, destroy cities, and even penetrate inland through amphibious or air assaults. However, without a heavy land component, naval or air threats, not to mention missiles, cannot occupy territory. In other words, the territory is vulnerable to an attack but not an invasion. This is not to underestimate the threat of missiles and the need to build a missile defense system, but this is a qualitatively different threat from an invasion and occupation of a state's territory.

A useful semantic and conceptual distinction is between *boundary* and *frontier.* A boundary is a defensive line separating the territorial sovereignty of one state from that of other states. It is the last line of defense. A frontier has a larger geographical scope and encompasses areas that are outside of the state sovereignty but vital to its security. For example, the United States has land boundaries with Canada and Mexico and maritime boundaries on the Atlantic and Pacific oceans. Its frontiers, however, have been drawn in Europe and the Pacific since the first half of the twentieth century. I am concerned here only with boundaries and their impact on foreign policy. See Lamb, *Asian Frontiers,* 8–9; Kristof, "Nature of Frontiers and Boundaries"; and Spykman, "Frontiers, Security, and International Organization."

62. Harold Sprout, "Frontiers of Defense," 218.

63. Earlier some (e.g., Mahan) argued that it was sea-based coalition led by Great Britain had cut France off from the rest of the world and defeated Napoleon's Continental empire. This argument assumes that because of the commercial importance of the sea, the defeat of a land power can be achieved by depriving it of access to sea lanes, making a seaborne invasion unnecessary. But this interpretation assigns too much value to sea power because ultimately the defeat of Napoleon was achieved on land, with Russia pushing toward France's territory. See Crowl, "Alfred Thayer Mahan," 452–54.

64. See also Starr and Most, "Substance and Study"; Paul Diehl, "Contiguity and Military Escalation"; Pearson, "Geographic Proximity"; Boulding, *Conflict and Defense,* 245–47; and Walt, *Origins of Alliances,* 23.

65. Spykman, "Frontiers, Security, and International Organization," 438.

66. I do not argue, therefore, that the geographic nature of borders is in itself a cause of conflict or peace. In other words, good natural fences do not necessarily make good neighbors, and bad fences do not necessarily make bad neighbors. The stability of borders depends on the equilibrium of power between the states separated by them. I agree with Spykman, who argued that the situation of borders is a reflection of the balance of power. "The position of that line [borders] may become an index to the power relations of the contending forces. Stability then suggests an approximation to balanced power, and shift indicate changes in the relative strength of the neighbors. . . . [Hence] the problem of territorial security becomes a problem of neutralizing power differentials." And elsewhere he writes that "interest in the frontier is now no longer in terms of the strategic value of the border zone but in terms of the power potential of the territory it surrounds." This argument reaffirms the fundamental point of this project: geography does not determine the political actions of states, but it serves as a background that influences them. See ibid., 437, 444.

67. Mahan seems to ignore the political characteristics of land borders and equates them with the inability to conduct a maritime geostrategy. Mahan, *Influence of Sea Power upon History,* 29. Hence, the English island has an unbridgeable advantage over Holland and France, which have one continental boundary, because it can pursue a maritime geostrategy, while the other two cannot. The assumption behind this argument, however, is that land borders simply because of their geographic nature require a greater expenditure of resources, hindering the state's ability to build a powerful navy. Again, I think such an argument ignores the political features of land borders that can be perfectly stable and secure, such as those of the United States.

68. See Desch, *When the Third World Matters,* 46–88.

69. See Mahan, *Influence of Sea Power upon History,* 25–27.

70. For instance, Robert Ross argued that China is unlikely to expand toward the Pacific not only because of its long continental frontier but also because of a strong "land culture." According to him, China has never developed an influential "sea culture" that could spur it to expand toward the sea. In other words, China has neither the territorial security neces-

sary to expand its sea power nor the motivation to do so. However, this argument hinges on a static view of borders because, as I have already mentioned, land borders are not automatically unstable. A benign border situation determined by a weak Russia, a friendly Central Asia, and a stable India gives China the possibility of becoming a Pacific Ocean power, vying to secure resources and vital routes. See Ross, "Geography of Peace," esp. 93-94, 103–8.

Three • The Geopolitical Change of the Sixteenth Century

1. Quoted in Scammell, "After Da Gama," 543.
2. Parry, *Spanish Seaborne Empire*, 117.
3. Rossabi, "'Decline' of the Central Asian Caravan Trade," 367.
4. Scammell, "After Da Gama," 520.
5. "The rise of Antwerp, Amsterdam, and London, and the contemporaneous decline of the Mediterranean both follow hard on the heels of the great explorations and the creation of early colonial empires. This lends chronological buttressing to the causal bridge between the Age of Discovery and the commercial and industrial burgeoning of northern Europe." Rapp, "Unmaking of the Mediterranean Trade Hegemony," 500. Rapp then proposes a slightly different interpretation of the decline of the Mediterranean: that it did not decline simply because Atlantic Europe had access to America and Asia but because it was penetrated by England and the Netherlands, who upset the regional hegemonies of Venice and the Ottomans. "The seventeenth century, especially the years 1620–1660, saw the establishment of the Atlantic hegemony. Although the victory was to the North, the theater of economic war was the Mediterranean market" (502). I do not disagree with this interpretation, which is shared by other historians, including, to a certain extent, Braudel. But the decline of the Mediterranean started much earlier, and the economic wars of the seventeenth century simply sealed the already irreparably weakened Venetian and Ottomans powers. See Rapp, "Unmaking of the Mediterranean Trade Hegemony"; and Lane, "Venetian Shipping during the Commercial Revolution."
6. See Panikkar, *Asia and Western Dominance*.
7. See van der Wee, "European Long-Distance Trade." See also Paul Kennedy, *Rise and Fall of British Naval Mastery*, 17–19.
8. The best example of this route connecting America with East Asia was the silver trade. In the late sixteenth century Manila became a key Spanish entrepôt of silver, bringing the metal from Potosi (Peru) to China. It was the birth of global trade. See Flynn and Giraldez, "Spanish Profitability in the Pacific"; Flynn, "Comparing the Tokagawa Shogunate with Hapsburg Spain"; and Dubbs and Smith, "Chinese in Mexico City."
9. For a fascinating description of this change and of how the various economic regions or systems of Eurasia became interconnected see also Abu-Lughod, *Before European Hegemony*.
10. Ma, "Great Silk Exchange," 60.
11. McNeill, "World History," 231. Elsewhere McNeill writes that "world history since 1500 may be thought of as a race between the West's growing power to molest the rest of the world and the increasingly desperate efforts of other peoples to stave Westerners off, either

by clinging more strenuously than before to their peculiar cultural inheritance or, when that failed, by appropriating aspects of Western civilization—especially technology—in the hope of thereby finding means to preserve their local autonomy." McNeill, *Rise of the West*, 652.

12. See, e.g., Bernard Lewis, *Muslim Discovery of Europe*, 48.

13. Scammell, "After Da Gama," 519.

14. Some argue that the geopolitical change of the fifteenth and sixteenth centuries was not as momentous as it is has been portrayed to be. According to this view, the Portuguese (and the Spaniards and Dutch that followed) simply inserted themselves into an already existing system of trade that linked Europe and Asia. Long-distance trade existed well before AD 1500, and the discovery of new maritime routes did not change the history of Eurasia or, in McNeill's words, of the "Eurasian ecumene." The term *ecumene* suggests that historical change in one region occurred in conjunction with or in response to changes in other regions. So-called regional histories were not immune to influences from other regions, despite the lack of cheap and fast communications. Hence, there was not a cataclysmic change in the 1500s because even before that "world history" had been moved by a series of interactions between the different, even very distant, societies. I do not disagree with such an interpretation insofar as it points to the pre-1500 existence of contacts between Asia and Europe. But I argue that the locale of convergence of these societies, of Europe and Asia, shifted from the center to the outside of Eurasia, and this change not only altered the location of the main routes between Europe and Asia but also introduced Europe to Asia. The meeting point between Europe and Asia was no longer in a neutral territory—in the Middle East or Central Asia—but directly in East Asia. This was a dramatic change from the pre-1500 world. See McNeill, "Changing Shape of World History," 11; idem, "World History," 220–21; Christian, "Silk Roads or Steppe Roads?" Bentley, "Hemispheric Integration"; Chaudhuri, *Asia before Europe*; Frank, *ReOrient*; and Subrahmanyam, "Connected Histories."

15. Describing the philosophy of Fernand Braudel (and his *Annales*), H. R. Trevor-Roper wrote, "There is the conviction that history is at least partly determined by forces which are external to man and yet not entirely neuter or independent of him, nor, for that matter, of each other: forces which are partly physical, visible, unchanging, or at least viscous and slow to change, like geography and climate, partly intangible, only intellectually perceptible, and more volatile, such as social formations or intellectual traditions." Trevor-Roper, "Fernand Braudel," 470–71.

16. Braudel observed that although the news of Vasco da Gama's voyage provoked panic in Venice, it did not result in the immediate switch from the Mediterranean to the Atlantic as the main link to Asia. In fact, the Mediterranean continued to prosper as a supply route of Asian goods through the end of the sixteenth century. See Braudel, *Mediterranean and the Mediterranean World*, 1:543–70.

17. The literature on this subject is massive and inconclusive. For a sample see Landes, *Wealth and Poverty of Nations*; Pomeranz, *Great Divergence*; Buck, "Was It Pluck or Luck?"; Vries, "Are Coal and Colonies Really Crucial?"; Landes, "What Room for Accident in History?"; Gale Stokes, "Fates of Human Societies"; idem, "Why the West?"; Dane Kennedy, "Expansion of Europe"; and Scammell, *First Imperial Age*.

18. Eric Jones, "Disasters and Economic Differentiation," 677.

19. The comparison between Atlantic Europe (or Europe) and China is always highly speculative because the two powers (or rather, China and some of the western European powers) did not meet in a war in the fifteenth through seventeenth centuries. By the time the Portuguese, and then the Spanish and Dutch, arrived in East Asia, China had already withdrawn from the world.

20. Deng, "Critical Survey," 22.

21. Parker, "Europe and the Wider World"; idem, *Military Revolution;* William Thompson, "Military Superiority Thesis"; Black, *War and the World.* See also A. R. Hall, "Epilogue"; and Howard, "Military Factor in European Expansion." Of course the technological explanation requires further explanation: why was European technology superior to that of the rest of the world? There is no end to the study of causes.

22. McNeill argues that Atlantic Europe had three "talismans of power" that gave it a clear advantage over the rest of the world: (1) a deep-rooted pugnacity and recklessness operating by means of (2) a complex military technology, most notably in naval matters; and (3) a population inured to a variety of diseases which had long been endemic throughout the Old World ecumene." McNeill, *Rise of the West,* 569.

23. Ibid., 570. See also William Thompson, "Military Superiority Thesis," 169.

24. It is also debatable whether the Europeans were technologically superior to Asian powers. Artillery was widely available, and the Portuguese did not have a clear advantage in this field.

25. Parker, "Europe and the Wider World," 194.

26. Marshall, "Western Arms in Maritime Asia."

27. Parker, "Europe and the Wider World," 179. It is significant that in Asia and in America defensive fortresses were erected mostly by Europeans when they had to prepare for war against other Europeans. "The native peoples of the Americas, Siberia and the Philippines, although they succeeded in emulating many military techniques of the European invaders, normally failed to build artillery fortresses." Idem, *Success Is Never Final,* 208.

28. Lane, "Economic Meaning of the Invention of the Compass."

29. Black, *Maps and History,* 6.

30. Parker, *Success Is Never Final,* 98, 120. Bernard Lewis writes that in the 1730s the Ottomans started to realize the importance of maps. In an influential book printed in 1731 an Ottoman official "discusses the value of scientific geography, the key to the knowledge of one's own and one's neighbors' territories, as a necessary part of the military art and as an aid to administration." Lewis, *Muslim Discovery of Europe,* 49.

31. A large galley capable of transporting a two-hundred-ton cargo required about two hundred men. Parry, *Age of Reconnaissance,* 69–71. For a very good description of the rise of Atlantic shipbuilding see ibid., 69–84.

Four • *The Geostrategy of Venice (1000–1600)*

1. From the mid-eleventh century until the loss of independence in 1797, each year the doge would sail out of the lagoons and in a symbolic gesture throw a golden ring into the sea. The formula "Sea, we wed thee in token of our true and perpetual dominion over thee" was regarded as a serious political commitment. See also Okey, *Story of Venice,* 64.

2. See also Grubb, "When Myths Lose Power"; and Benzoni, "Venezia."

3. Brown, "Venice," 413–14; Hodgson, *Early History of Venice*, 317.

4. According to A. T. Mahan, "The feature which the steamer and the galley have in common is the ability to move in any direction independent of the wind. Such a power makes a radical distinction between those classes of vessels and the sailing-ship; for the latter can follow only a limited number of courses when the wind blows, and must remain motionless when it fails." Mahan, *Influence of Sea Power upon History*, 3.

5. Lane, *Venice*, 248.

6. Economic resources are those produced through human effort; in the case of Venice these were Asian spices and silk. Natural resources are found rather than produced, for example, wood or water, both vital for Venice.

7. For an exhaustive study of the politics of salt in Venice see Hocquet, *Le sel et la fortune de Venise*.

8. Lane, *Venice*, 58.

9. The strategic importance of timber was well illustrated by the pontifical injunction compelling Christians not to sell it to the Saracens in the tenth and eleventh centuries. The objective was to limit Muslim military capabilities. The Venetians, however, rarely obeyed ecclesiastical or imperial orders that conflicted with their commercial interests and used their supply of timber to develop commercial relations. For the strategic importance of timber and the supply difficulties see also McNeill, *Venice: The Hinge of Europe*, 145–46; and Hodgson, *Early History of Venice*, 164, 249.

10. See Norwich, *Venice: The Rise to Empire*, 109.

11. For a detailed examination of how timber shortages affected Venetian shipbuilding see Lane, "Venetian Shipping during the Commercial Revolution," esp. 224.

12. In a 1910 article J. Russell Smith argued that Venice was a "world entrepôt," distributing certain products to a whole region, because the goods it traded had "high value and small bulk," and the origins (Asia) and the destination (Europe) of the trade were distant from each other. "The more remote the origins and destinations of the traffic, the stronger is the hold upon this trade of the entrepôt with its organization of routes." With changes in routes and in the features of the traded goods, Venice lost its role to England. See Smith, "World Entrepôt."

13. Chalandon, "Earlier Comneni," 325.

14. The rise of the Ottoman Empire is often considered to be one reason why the Portuguese and the Spanish attempted to discover a new route to India: in order to avoid Muslim territories. While I do not wish to enter this debate regarding Portugal and Spain, it seems clear that Venice did not have a similar incentive to seek new avenues for trade and happily continued to use the well-known Mediterranean routes. The rise of the Ottomans did not change the geopolitical reality faced by Venice.

15. See Bloch, *Feudal Society*, 70–71.

16. For a classic book see Haskins, *Renaissance of the Twelfth Century*. On the effects of Europe's growth see also Gibbon, *History of the Decline and Fall of the Roman Empire*, 3:669.

17. On the role of cities in medieval commerce see also Pirenne, *Medieval Cities*.

18. On the national developments of the "reconcentration of authority" see Bloch, *Feudal Society*, 421–37.

19. For instance, the power of France was so great that in 1309 the French king, Philippe le Bel, convinced the pope to move to Avignon, thus initiating the so-called Babylonian captivity of the papacy, which lasted until 1377.

20. Luzzati, "La dinamica secolare," 27.

21. For a brief overview of direct contacts between Europe and Asia see Lopez, "European Merchants in the Medieval Indies." From the time of the Roman Empire the Mediterranean had been a conduit for commercial and political relations among the communities nestled on its shores. The early medieval decline of interactions on this sea proved to be temporary, and by the eleventh century the Mediterranean was again becoming an avenue of commerce.

22. Okey, *Story of Venice*, 91.

23. See also Karpov, *L'impero di Trebisonda*, 32–34.

24. Riley-Smith, *Crusades*, 203.

25. Geographically Alexandria was a difficult port because winds blew from the northwest, easing the arrival of ships from Europe, only during the summer months. Departure was virtually impossible until the winter, when winds turned westward but made navigation dangerous owing to violent storms. See Lane, *Venice*, 73.

26. "It is probable that the greater readiness of Italian, as compared with the Byzantine traders to enter into commercial relations with unbelievers was the main cause of the gradual transfer of the carrying trade of the Mediterranean from Byzantine to Italian hands." Hodgson, *Early History of Venice*, 160–61. Venice traded with Muslim Alexandria as early as the eighth century. For instance, in 828 Venetian merchants on a commercial trip to Alexandria stole St. Mark's relics from the Egyptian port. See Charles Diehl, *La République de Venise*, 38.

27. See also Hoffmann, "Commerce of the German Alpine Passes"; and an interesting analysis of land routes using numismatic data, Adelson, "Early Medieval Trade Routes." Political changes during the fourteenth and fifteenth centuries, in particular the Hundred Years' War, favored the economic development of Central Europe to the detriment of France. Such a shift diminished the importance of ports connected with French markets, such as Genoa, and gave greater leverage to Venice, the harbor of choice for German towns. See also van der Wee, "European Long-Distance Trade," 20–21.

28. The strength of Amalfi was that it traded with the Arabs of North Africa, supplying them with key resources, such as timber. Such regional commerce limited Amalfi's reach in the eastern Mediterranean. Citarella, "Patterns in Medieval Trade."

29. Pryor, *Geography, Technology, and War*, 93–94.

30. For a detailed geographic description of Venice see East, *Historical Geography of Europe*, 304–8.

31. Gibbon, *History of the Decline and Fall of the Roman Empire*, 2:345.

32. Lane, *Venice*, 193–94. For a description of the role of the lagoon pilots see ibid., 17–18.

33. Ibid., 225.

34. A short note on the choice of the historical period under consideration is necessary. Venetian history spans more than ten centuries, starting in the midst of legends dating back to the sixth century, when refugees from northern Italy, escaping from the violence of

the Huns, settled on the geologically precarious but strategically safe lagoons. It ends in 1797 with the Treaty of Campo Formio, through which Napoleon ceded Venice to Austria. However, the period 1600–1500 represents the peak of Venetian power. Before the eleventh century Venice limited its influence over local trade and river navigation. And after the beginning of the sixteenth century, with the 1509 League of Cambrai, Venice became a second-rate power that, albeit not retreating from its strategic dominions, had little leverage over the course of European history.

35. Machiavelli succinctly described the reasons for Venetian success. He wrote: "It is said that the Venetians in all those places which they are recovering are painting a lion of St. Mark which has in its hand a sword rather than a book, from which it seems that they have learnt to their cost that study and books are not sufficient to defend states." Niccolò Machiavelli, quoted in Finlay, "Fabius Maximus in Venice," 998.

36. See also Ravegnani, "Tra i due imperi."

37. Even if under Venetian hegemony since the thirteenth century, Ferrara was allowed to conduct its commerce with the rest of Italy and Europe freely and without extensive Venetian control. See Dean, "Venetian Economic Hegemony."

38. See also Pertusi, *Saggi veneto-bizantini*, 56 and passim.

39. Brown, "Venice," 405.

40. On the wine see Thayer, *Short History of Venice*, 35. Venice's reluctance to use violence to conquer territories, preferring diplomacy and commercial sanctions to achieve limited economic and political goals, was probably also owing to the delicate political situation of the lagoons. The Western Empire and the pope were too powerful to be ignored, and their potential response to a violent expansion on the part of Venice could have jeopardized the commercial standing of Venice in Italy. Therefore, the Venetian doges were careful not to spark a counterbalancing action by these mainland powers.

41. Brown, "Venice," 401.

42. Bréhier, "Greek Church," 273; idem, "Attempts at Reunion," 597.

43. Pertusi, *Saggi veneto-bizantini*, 60.

44. It is unclear when the first *promissio* was formulated. In 1192 Enrico Dandolo pronounced the first official *promissio*, but there are indications that the preceding doges used similar formulas. See ibid., 118–19.

45. Hodgson, *Early History of Venice*, 327.

46. Madden, "Venice's Hostage Crisis."

47. Queller and Day, "Some Arguments in Defense of the Venetians," 724.

48. Queller, *Fourth Crusade*, 50–51.

49. See, e.g., Charles Diehl, "Fourth Crusade and the Latin Empire," 416–17.

50. Godfrey, *1204: The Unholy Crusade*, 9–10, 24–28.

51. Queller and Day, "Some Arguments in Defense of the Venetians," 718.

52. Bravetta, *Enrico Dandolo*, 18.

53. For instance, by leading the early crusades Genoa obtained trading privileges in the Kingdom of Jerusalem (1101 and 1104) and in the Principality of Antioch (1101). Queller and Katele, "Venice and the Conquest," 26.

54. It has also been argued that Enrico Dandolo had a strong personal animosity toward Byzantium. According to more a legend than a proven fact, during the 1171 hostage crisis

Dandolo was blinded by the Byzantines, who considered him a dangerous "enemy of the Empire." See Bravetta, *Enrico Dandolo*, 22–23.

55. Perhaps illustrative of the Venetian strategic thinking, while the Western crusaders were satisfied by random pillaging, burning, and destroying, the Venetians carefully chose their booty. The four bronze horses standing in front of the St. Mark's Basilica in Venice were among the spoils.

56. Interestingly, his wife followed him in 1204 but, unaware of the diversion of the crusade, sailed to Acre in the Holy Land. The diversion of the crusade was not known in Europe until after the final conquest of Constantinople. Godfrey, *1204: The Unholy Crusade*, 135.

57. Some historians argue that Constantinople was the nucleus of the Venetian empire; see e.g., Lane, *Venice*, 43. While it was certainly a key possession, it did not constitute the cornerstone of Venetian power. Even after the loss of the city in 1262, Venice continued to be the dominant power in the Mediterranean.

58. For a description of the debate see Thayer, *Short History of Venice*, 76–78.

59. McNeill, *Venice: The Hinge of Europe*, 32–33.

60. Lane, *Venice*, 43.

61. On the navigational importance of Crete see Pryor, *Geography, Technology, and War*, 94–95.

62. On the conquests of Dandolo's nephews and their strategic importance see Ravegnani, "La Romania veneziana," 199–200.

63. Pryor, *Geography, Technology, and War*, 101.

64. Although in many instances Genoese ships' crews refused to fight, according to some these battles constituted the beginning of real naval strategy and tactics. See Devries, *Medieval Military Technology*, 198.

65. See also Katele, "Piracy and the Venetian State."

66. The voyages of Marco Polo illustrate the extent of Venice's penetration of the Black Sea. At the end of the thirteenth century Marco Polo traveled through the Black Sea to China, demonstrating the profitability of this route.

67. Because of adverse currents, ships often had to seek harbor at Tenedos, waiting several days, sometimes weeks, for winds to shift in their favor. See Pryor, *Geography, Technology, and War*, 98.

68. See Mallett and Hale, *Military Organization of a Renaissance State*, 5; and Coccon, *La Venezia di Terra*, 11–13.

69. See also Zakythinos, "L'attitude de Venise," 61–65.

70. Lane, "Recent Studies," 322. Lane continues: "The immediate effects of the rise of the Ottoman empire on Venetian trade have generally been exaggerated, particularly in the United States. Writers of textbooks have found it convenient to connect the history of the Old World and the New by saying that the discovery of America was caused by the Turks having blocked the routes to the east. . . . A belief that the Turks would necessarily have a devastating effect wherever they went was fostered by their nineteenth-century reputation as the 'terrible Turk.'" Ibid.

71. The Ottoman Empire did not deprive Venice of its main bases before the end of the sixteenth century: Cyprus in 1571, Crete in 1669, and Morea in 1716.

72. See Costantini, *L'acqua di Venezia*, 14–15.

73. Okey, *Story of Venice*, 129. On the advantages of the Brenta as compared with other rivers see Costantini, *L'acqua di Venezia*, 28.

74. Coccon, *La Venezia di Terra*, 11. An alternative explanation for Venetian involvement on the Italian mainland asserts that the increasingly powerful Venetian nobility preferred an easy conquest of nearby territories, avoiding the increasingly more dangerous maritime bases, threatened by the Ottomans. In other words, the pressure of the Ottoman power in the East pushed Venice—or rather, convinced Venetian nobles—to seek profits on the nearby Italian peninsula. The explanation is not convincing for two reasons. First, Venice continued a profitable commerce in the East, despite the Ottoman pressures. In fact, as we have seen, Venice started to lose ground to the Ottomans about 1470, after its main thrust toward the *terraferma*. Second, the involvement of the Venetians, through large investments in property, was the result, rather than the cause, of Venice's expansion in Italy. The bulk of money went to the mainland only in the sixteenth and seventeenth centuries, well after the conquest of the Italian region by Venice. See also Luzzatto, *Storia economica di Venezia*, 147–48.

75. Baron, "Struggle for Liberty in the Renaissance," 274–75.

76. Okey, *Story of Venice*, 164–65; Lane, *Venice*, 227. The famous justification for the murder of the Carraras was that "uomo morto no fa guerra" (dead men wage no wars).

77. Venice's ease in expanding on the *terraferma*, according to Machiavelli, was owing to the lack of liberty in the neighboring cities. "Though Venice's neighbours are more powerful than those of Florence, yet, on account of its having found the cities less obstinate, Venice has been able to subdue them more quickly than has Florence, which is surrounded entirely by free cities." Niccolò Machiavelli, *Discorsi*, bk. 3, chap. 12, in Machiavelli, *Il Principe e altre opere politiche*, 386, my translation. The geopolitical location of Venice was, therefore, superior to that of Florence. Venice's expansion to the mainland altered that.

78. Baron, "Anti-Florentine Discourse," 332.

79. Baron, "Struggle for Liberty in the Renaissance," 562.

80. See also Mallett and Hale, *Military Organization of a Renaissance State*, 181–98.

81. Carmagnola was put on trial for treason and beheaded.

82. For more information on the size of the river fleets see Mallett and Hale, *Military Organization of a Renaissance State*, 96–100.

83. Lane, *Venice*, 231.

84. For the costs of the expansion on the *terraferma* see Luzzatto, *Storia economica di Venezia*, 163–64.

85. Fubini, "Italian League," 174.

86. Okey, *Story of Venice*, 185. Sforza's warning is strikingly similar to that of a Spanish diplomat to Philip II in 1600: "Truly, sir, I believe we are gradually becoming the target at which the whole world wants to shoot its arrows; and you know that no empire, however great, has been able to sustain many wars in different areas for long. If we can think only of defending ourselves, and never manage to contrive a great offensive blow against one of our enemies, so that when that is over we can turn to the others, although I may be mistaken I doubt whether we can sustain an empire as scattered as ours." Parker, *Grand Strategy of Philip II*, 111.

87. Guicciardini, *Storie fiorentine*, 135.

88. See, e.g., Coccon, *La Venezia di Terra*, 48–49.

89. Niccolò Machiavelli, *Il Principe*, chap. 21, in Machiavelli, *Il Principe e altre opere politiche*, 85, my translation.

90. More exactly, the two Italian powers with the greatest ambition were Venice and the pope. However, the pope could be controlled by internal divisions in Rome, while "to keep the Venetians back, the union of all other Italian powers was necessary." See ibid., chap. 11, p. 48, my translation.

91. Finlay, "Fabius Maximus in Venice," 994.

92. On the complex diplomacy of this event, particularly between Venice and the papacy, see Gilbert, *The Pope, His Banker, and Venice*. See also Finlay, "Fabius Maximus in Venice," 1025.

93. Venice was, in McNeill's words, "a marginal polity, balanced precariously between Ottoman east and Christian west, and caught no less precariously between the Austrian and Spanish centers of Hapsburg power, the one pressing down from the north, the other reaching up from the south." McNeill, *Venice: The Hinge of Europe*, 126.

94. "Venice, in later times, figured more than once in wars of ambition; 'till becoming an object to the other Italian States, Pope Julius the Second found means to accomplish that formidable league [of Cambrai], which gave a deadly blow to the power and pride of this haughty republic." Hamilton, Madison, and Jay, *Federalist Papers*, no. 6, pp. 24–25.

95. Machiavelli, *Discorsi*, bk. 3, chap. 11, p. 384, my translation. After the 1470 fall of Negroponte some Venetians perceived the danger in having failed to defend their maritime possession. A Milanese ambassador in Venice observed that after the news of the Ottoman conquest reached the city, "all Venice is in the grip of horror: the inhabitants, half dead with fear, are saying that to give up all their possessions on the mainland would have been a lesser evil." But once people calmed down, Venice continued its mainland policy. See Babinger, *Mehmed the Conqueror and His Time*, 284.

96. Ranke defends Venice's decision to expand in the Italian mainland because the "sea possessions were unable to support themselves." Venice needed the food and the financial wealth from the northern Italian cities to maintain its imperial framework. But Ranke's argument is based on the situation at the end of the sixteenth century, when Venice already had vast territories in Italy. The poverty of its sea possessions was in part owing to Venice's neglect of them a century earlier. See Ranke, *Venezia nel Cinquecento*, 96–97.

97. Thayer, *Short History of Venice*, 207.

98. The news of Columbus's voyage spurred only curiosity in Venice. Venetian envoys obtained copies of Columbus's charts in Spain after paying him. See Okey, *Story of Venice*, 191.

99. Luzzatto, *Storia economica di Venezia*, 219.

100. Smith, "World Entrepôt."

101. Quoted in Finlay, "Crisis and Crusade," 57.

102. In fact, some Venetians advocated ending the conflict with the Ottomans in order to concentrate on the Portuguese threat. Girolamo Priuli wrote: "This [the Portuguese discovery of the Cape route] is more important to the Venetian state than the war with the Turks or any other war that could take place. . . . And if this route continues—and already it

appears to me easy to accomplish—the king of Portugal might be called the king of money." Quoted in Finlay, "Crisis and Crusade," 57.

103. Lane, *Venice,* 293.

104. See Lane, "Pepper Prices before Da Gama," 596–97.

105. It has also been argued that the shift of European power from the Mediterranean to the Atlantic shores was owing to their rapid industrial development rather than to geographic discoveries. It was "the invasion of the Mediterranean, not the exploitation of the Atlantic, that produced the Golden Age of Amsterdam and London." Rapp, "Unmaking of the Mediterranean Trade Hegemony," 501. This seems to be a chicken-and-egg problem. In my opinion, it is difficult to detach the discovery of new routes and continents from the rise of the Atlantic regions. The new powers never fully replaced Venice in the Mediterranean (until its defeat by Napoleon at the end of the eighteenth century); they merely circumvented its imperial power by exploiting new routes. See ibid., 499–525.

106. See Verlinden, "Venise entre Méditerranée et Atlantique."

107. For a fascinating albeit often esoteric description of navigation on the Mediterranean Sea see Guilmartin, *Gunpowder and Galleys,* esp. 57–84.

108. For the role of geography in military change see Downing, *Military Revolution and Political Change,* 79.

109. Burckhardt, *Civilization of the Renaissance in Italy,* 40.

110. Godfrey, *1204: The Unholy Crusade,* 57.

111. Thayer, *Short History of Venice,* 198.

Five • The Geostrategy of the Ottoman Empire (1300–1699)

1. Jan Sobieski to Marysieńka Sobieska, 13 September 1683, in Sobieski, *Jan Sobieski: Listy do Marysieńki,* 522.

2. Inalcik, *Essays in Ottoman History,* 20.

3. For an interesting summary of the "straits question" see Hurewitz, "Russia and the Turkish Straits"; and Jelavich, *Ottoman Empire.*

4. The shortage of studies on the origins of the Ottoman Empire is owing in part to the fact that the first decades and even centuries of the Ottomans' rise to power attracted scant attention in Europe. Their origins remained shrouded in mystery and legend. Even the classic academic text on the early decades of the Ottomans is emphasizes the mythical origins of the Ottoman "race." See Gibbons, *Foundation of the Ottoman Empire.* The Ottoman historiography, on the other hand, tends to portray the early years in legendary terms, producing works closer to epic than history. See Köprülü, *Origins of the Ottoman Empire,* 3–21. Also, because of the European fear of a Turkish flood, since the Renaissance there has been a strong bias against the Ottomans. Schwoebel, "Coexistence," 165.

5. Gibbons, *Foundation of the Ottoman Empire;* Paul Wittek, *Rise of the Ottoman Empire.*

6. Mahan, *Influence of Sea Power upon History,* 29.

7. See Doyle, *Empires,* 106–7.

8. Paul Wittek, *Rise of the Ottoman Empire,* 2. For a summary of the historiography see Kafadar, *Between Two Worlds,* 9–12 and esp. 29–59; and Langer and Blake, "Rise of the Ottoman Turks."

9. For instance, as Gibbons points out, Osman was a "fanatic, if by fanatic is meant one who is stirred by religious zeal and makes his religion the first and prime object in his life." Gibbons, *Foundation of the Ottoman Empire*, 53. But he and his successors were not intolerant and allowed Christians to live and prosper under their rule. See also ibid., 73–81.

10. Thuasne, *Gentile Bellini et Sultan Mohammed II*, 27, my translation. See also Cot, *Suleiman the Magnificent*, 201–3. There are many examples of sultans maintaining more than friendly relationship with Christian leaders. For instance, Muhammed the Conqueror held Lorenzo de' Medici in such high esteem that when, after the Congiura de' Pazzi, a conspiracy led by the Pazzi family against the ruling Medici family, in Florence, one of its instigators sought refuge in Constantinople, the sultan arrested the man and turned him over to a Florentine ambassador. See Thuasne, *Gentile Bellini et Sultan Mohammed II*, 43–45.

11. Thuasne, *Gentile Bellini et Sultan Mohammed II*, 10.

12. Paul Wittek, *Rise of the Ottoman Empire*, 49.

13. On the education and training of the Janissaries see Goodwin, *Janissaries*, 32–53.

14. Horniker, "Corps of the Janizaries," 178.

15. Nomadic warfare "works well in the steppe, when problems of supply prevent determined pursuit by a sedentary power, or in mountainous areas." Lindner, *Nomads and Ottomans*, 31.

16. See ibid., 3–5.

17. See, e.g., Wittek's comments regarding the situation in Anatolia after Tamerlane's invasion. At that time the Anatolian princes again became independent, forcing the Ottomans to redirect their efforts toward the Balkans. By implication, had the princes been weak and disunited, the Ottomans would have abandoned the *ghaza* in Europe and pursued territorial expansion in Anatolia. Paul Wittek, *Rise of the Ottoman Empire*, 48.

18. Goodwin, *Janissaries*, 25. Suleiman the Magnificent, for instance, while a devout Muslim, allowed freedom of religion in the Ottoman Empire. At the same time, however, he pursued a repressive policy toward the Shiites. See Cot, *Suleiman the Magnificent*, 71–72.

19. See Gilpin, *War and Change*, 37; Wolfers, "Pole of Power and the Pole of Indifference"; and Rosecrance, *International Relations*, 141–43.

20. Ranke, *Ottoman and the Spanish Empires*, 1.

21. Langer and Blake, "Rise of the Ottoman Turks," 476. They continued: "The salt sink in the center, to the west of the Halys River, impedes direct longitudinal traffic; the wooded areas on the coast have little in common and no connection with the barren plains of the uplands; and the deep river valleys are effectually sundered from each other by the mountains. . . . These individual and sharply marked geographical units form the districts and cantons which play a large part in the life of Asia Minor." Ibid.

22. See Köprülü, *Origins of the Ottoman Empire*, 111.

23. Potkowski, *Warna 1444*, 18. In the words of the historian Edwin Pears, "The greater body was constantly attracting to itself members of the smaller bodies." Pears, "Ottoman Turks," 668. The natural wealth of this region and the proximity of even greater wealth encouraged the migration of people even from Central Asia, enlarging the Ottoman ranks. As Langer and Blake observe, "The movement from the dry, barren uplands to the thicker vegetation of the river valleys and the better grazing areas on the north side of Olympus and

ultimately to the richer, busier lowlands about Bursa and Yenišehir was the most natural thing in the world." Langer and Blake, "Rise of the Ottoman Turks," 494–95.

24. McCarthy, *Ottoman Turks*, 38.

25. Köprülü, *Origins of the Ottoman Empire*, 47. The rapid and large increase in population also allowed these tribes to take huge losses. See also Pears, "Ottoman Turks," 655.

26. "The rapidity with which the Ottomans entered Europe and became firmly established in Gallipoli, after easily taking an important part of Karasid territory, was a major factor in strengthening the structure of their state, because a great many nomadic elements, poor villagers and sipahis, who wanted to acquire rich timars in Rumelia, went from central Anatolia and . . . coastal beyliks . . . to Thrace and Macedonia in order to find and settle rich empty lands." Köprülü, *Origins of the Ottoman Empire*, 114.

27. Madden, *Concise History of the Crusades*, 195.

28. Paul Wittek, *Rise of the Ottoman Empire*, 32.

29. Often the opposite happened, that is, European soldiers joined Turkish groups. For instance, in 1305 a detachment of Catalan soldiers joined some Turkish fighters. The descendants of these Christian renegades (e.g., the Mikhaloghlu-Michaelsons family) enjoyed great fame and success under Ottoman rulers. See ibid., 42; and Goodwin, *Janissaries*, 27, 33.

30. Paul Wittek, *Rise of the Ottoman Empire*, 44.

31. Pears, "Ottoman Turks," 662. There is some disagreement concerning the date of the fall of Nicea. The year 1329 seems to be the most widely accepted, although some historians mention the exact date 2 March 1331. See Shaw, *History of the Ottoman Empire*, 15; and Bryer, "Nicaea, a Byzantine City."

32. Pears, "Ottoman Turks," 662.

33. Bryer, "Greek Historians on the Turks"; Gibbons, *Foundation of the Ottoman Empire*, 93–94.

34. Gibbons, *Foundation of the Ottoman Empire*, 108.

35. Potkowski, *Warna 1444*, 24.

36. Inalcik, *Ottoman Empire*, 9.

37. Ibid., 420.

38. Pears, "Ottoman Turks," 669.

39. Ibid., 667. By then the Ottoman "state had in its possession the equipment necessary for these expeditions, such as an army and an administrative executive." Paul Wittek, *Rise of the Ottoman Empire*, 45.

40. Pears, "Ottoman Turks," 667. Among other things, this invasion was a logistical marvel. Over the course of three nights more than thirty thousand Ottoman soldiers crossed the northern part of the Dardanelles, transported by ships and, apparently, crossing a bridge of inflated skins.

41. According to Halil Inalcik, "Geographic conditions determined the pattern of Ottoman conquest in the Balkans. They followed the direction of the historic Via Egnatia towards the west, reaching the Albanian coast in 1385, by way of Serres, Monastir and Okhrida. The local lords in Macedonia and Albania accepted Ottoman suzerainty. A second line of advance was against Thessaly, with the port and city of Salonica falling in 1387; a third followed the road from Constantinople to Belgrade and, in 1365, the Maritsa

valley came, with little resistance, under Ottoman control. . . . [By the late fourteenth century] the Ottomans controlled the main routes in the Balkan peninsula." Inalcik, *Ottoman Empire*, 11.

42. Potkowski, *Warna 1444*, 27.

43. Gibbons, *Foundation of the Ottoman Empire*, 122.

44. Ibid., 107.

45. Pears, "Ottoman Turks," 670. Venice was similarly friendly with the Ottomans. In 1375, during the war between Venice and Genoa, the Venetians were friendly toward the Ottoman power. For example, after Venetian troops occupied the strategic island of Tenedos, controlling the entrance to the Dardanelles, they were given the standing order to cede the fortress to the Ottomans rather than to the Genoese if they were unable to hold it.

46. Pope Gregory XI died in 1378, and the papacy became divided between Rome and Avignon. As a result, there was no attempt to stop the Ottomans in Europe. See Gibbons, *Foundation of the Ottoman Empire*, 137–38.

47. It should be noted that the Black Death wreaked havoc in Europe in 1346, with a serious recurrence in the 1360s. The decimated European populations had no energy to counter an Ottoman advance in Asia Minor. As Gibbons observed, the period of the Black Death and the related plagues "coincide with the most aggressive period of Ottoman conquest." Ibid., 96.

48. Pears, "Ottoman Turks," 670.

49. Gibbons, *Foundation of the Ottoman Empire*, 125, 126. Halil Inalcik has a slightly different interpretation. He argues that the Ottoman expansion was equally directed toward Europe and Asia, merely alternating between the two: "Ottoman advances in Europe were always paralleled by an expansion of their territory in Asia, an advance on one front following an advance on the other." This alternation, however, does not deny the strategic importance of the European conquests. In other words, Ottoman sultans could expand in Asia because of the resources obtained in Europe, but they could not expand in Europe, basing themselves only in resource-poor Asia Minor. Inalcik, *Ottoman Empire*, 14.

50. See McGowan, "The Middle Danube *cul-de-sac*," 170.

51. Pears, "Ottoman Turks," 675.

52. Participating in this battle were troops from all of the Balkan and eastern regions except for the soldiers of the Byzantine ruler. "As the battle on the Maritza had broken the power of the South Serbs and of the eastern Bulgarians in 1371, so did this battle on the plains of Kossovo in 1389 destroy that of the northern Serbians and the western Bulgarians." Ibid., 672. See also Potkowski, *Warna 1444*, 34.

53. There are several stories about how Murad lost his life. According to Serbian legends, Murad was killed by one Milosh Obravitch, who, accused of treason by his Serbian prince, was eager to prove his loyalty. He infiltrated the Ottoman ranks as a deserter and, after obtaining an audience with Murad, killed him with a dagger. "It is a commentary on the Serbian character that this questionable act has been held up to posterity as the most saintly and heroic deed of national history." Gibbons, *Foundation of the Ottoman Empire*, 177.

54. The French knights, partly because of their low opinion of the Ottoman soldiers and partly because of their dislike of the Hungarian infantry, charged the Ottoman ranks on

their own. In making their powerful charge they became separated from the Hungarians; deprived of the Hungarians' defensive infantry cover, the knights were decimated by the Ottomans soldiers. Pears, "Ottoman Turks," 675–76.

55. A fifteenth-century Polish historian, Jan Długosz, wrote that "this terrible defeat and this painful disgrace bestowed upon the Turks such strength and courage that they became more daring and more cruel, and became eager to achieve great things." Quoted in Potkowski, *Warna 1444*, 55, my translation.

56. Pears, "Ottoman Turks," 677.

57. "Including the stretch across Asia Minor from Angora to the Bosphorus, the total time for a journey from the eastern to the western boundaries of the Ottoman Empire was forty-two days. At the time of Murad's accession, twenty-seven years earlier, there was a mere three days' journey between them." Kinross, *Ottoman Centuries*, 55.

58. Inalcik, *Ottoman Empire*, 16.

59. Tamerlane died in 1405, before he was able to pursue his plans against Ming China.

60. Pears, "Ottoman Turks," 684.

61. Paul Wittek, *Rise of the Ottoman Empire*, 47.

62. Pears, "Ottoman Turks," 683.

63. As stated earlier, the opposite was probably unlikely. Had the Ottomans lost their European territories, the straits could have served as a line of defense, but the Asian territories could not have provided the wealth necessary to reconstruct the empire.

64. See also Pears, "Ottoman Turks," 687.

65. Inalcik, *Ottoman Empire*, 96.

66. Potkowski, *Warna 1444*, 123.

67. For example, in 1439 the Ottoman sultan sent an embassy to Cracow inviting the Polish king to join him in an anti-Habsburg alliance. Because the Poles had just signed a truce with the Habsburgs and were growing increasingly worried about the rising Ottoman power, the sultan's diplomatic mission failed. However, it showed how actively the Ottoman diplomats were attempting to prevent European unity. See ibid., 90–92.

68. Ibid., 99.

69. Babinger, *Mehmed the Conqueror and His Time*, 28. See also Potkowski, *Warna 1444*, 137.

70. Potkowski, *Warna 1444*, 143.

71. Ibid., 147–48.

72. Pears, "Ottoman Turks," 691; Potkowski, *Warna 1444*, 159.

73. Potkowski, *Warna 1444*, 161–62.

74. There is still an animated debate concerning how Ottoman forces crossed the Bosphorus. Venetian and Genoese galleys were supposed to control this stretch of water while the Hungarian, Polish, and other European land troops advanced from the north. In reality, an Ottoman army of forty thousand soldiers crossed the straits without too much difficulty. The explanations include a betrayal of the Venetians commanders, aid from the Byzantine emperor, and weather and sea currents that threw the European galleys off course. For an overview see Potkowski, *Warna 1444*, 182–84; for a version stressing the betrayal of the Genoese see Pajewski, *Buńczuk i koncerz*, 14–16.

75. Potkowski, *Warna 1444*, 198. The tactical blunder at Varna was remarkably similar to that committed by French kings at Nicopolis in 1396.

76. For example, in 1448 the Hungarians launched a new campaign against the Ottomans and encouraged the Albanian population to rebel. But inferior in number and without a ruling elite, the Hungarians were defeated again. See Swoboda, *Warna 1444*, 52–53.

77. Potkowski, *Warna 1444*, 210–11.

78. Pears, "Ottoman Turks," 693.

79. For a review of some sources on Muhammed the Conqueror see Inalcik, "Mehmed the Conqueror."

80. Pears, "Ottoman Turks," 694.

81. The Ottomans also had very friendly relations with Genoa and its merchants in the eastern Mediterranean. Karen Fleet writes that the "main motivating force behind these relations was money, generated by an active and lucrative commerce." The Ottomans' friendship with the Genoese continued even during the siege of Constantinople, when the Genoese "managed to maintain their relations with the Turks while, simultaneously, siding with the defenders of the city, sending letters urgently requesting help to Genoa, . . . soldiers to Constantinople, oil for Turkish cannons to the sultan's camp." Fleet, *European and Islamic Trade*, 11–12.

82. "The blood flowed in the city like rainwater in the gutters after a sudden storm, and the corpses of Turks and Christians were thrown into the Dardanelles, where they floated out to sea like melons along a canal." Barbaro, *Diary of the Siege*, 67. The siege of Constantinople was characterized by a huge disparity in numbers. About 150,000 Ottoman soldiers attacked the city, which was defended by about 8,000 defenders, mostly Europeans. Moreover, the Ottomans had a substantial technological advantage. They used artillery, notably a bombard cannon built by a Christian Hungarian that shot a twelve-hundred-pound ball seven times a day. See Pears, "Ottoman Turks," 697; Archibald Lewis, "Islamic World and the Latin West," 835; and McNeill, *Pursuit of Power*, 61.

83. Babinger, *Mehmed the Conqueror and His Time*, 98. See also Schwoebel, *Shadow of the Crescent*, esp. 1–81.

84. In 1459, in an attempt to stir an anti-Ottoman expedition, Pope Pius II inveighed: "We ourselves allowed Constantinople, the capital of the East, to be conquered by the Turks. And while we sit at home in ease and idleness, the arms of the barbarians are advancing to the Danube and the Sava. . . . All this [Ottoman conquests] happened beneath our very eyes, but we lie in deep sleep. No, we are able to fight among ourselves, but let the Turks do as they please. For trifling provocations the Christians take up arms and fight bloody battles; but against the Turks, who blaspheme our God, destroy our churches, and seek to extirpate the very name of Christianity, no one is willing to raise a hand." Despite this impassioned two-hour-long speech, the pope failed to persuade the Venetians, the only power with the necessary fleet, to organize an expedition against the Ottomans. Babinger, *Mehmed the Conqueror and His Time*, 170–71.

85. Pears, "Ottoman Turks," 705.

86. According to Fernand Braudel, the era of the Crusades ended only with Lepanto in 1571. But in the sixteenth century the crusading fervor was directed more against Protestant

Germany, where the Roman popes were more interested in combating the Protestant heresy than in checking Ottoman expansion. See Braudel, *Mediterranean and the Mediterranean World*, 2:842–84. Moreover, after the fall of Constantinople a growing literature in the West, including Italy, accorded the Ottomans legitimacy of power, reflecting the change in policy. With the news of the fall of Constantinople still fresh, Venice signed a commercial treaty with the sultan in 1454. A few years later, in 1479, the Venetians sent a friendly embassy led by Gentile Bellini to Sultan Muhammed II. See Schwoebel, "Coexistence," 164–87; and Thuasne, *Gentile Bellini et Sultan Mohammed II.*

87. Inalcik, *Ottoman Empire,* 26.

88. "The Golden Horn may have owed its name to the rich shoals of fish, notably the huge tunny, which moved along with the current from the Black Sea and constituted in the early days of the town its chief revenue; it was difficult of entry because of the north-east winds but afforded a deep sheltered roadstead which was navigable for seven miles inland." See East, *Historical Geography of Europe,* 184–85.

89. Cot, *Suleiman the Magnificent,* 200.

90. Until the conquest of Constantinople, Ottoman sea power was not taken seriously by European states. The sultan controlled some corsairs based in the Aegean, which, however, never engaged Christian powers in a naval battle. In 1423 a French captain considered Gallipoli, with only four galleys, to be the main Ottoman base. In 1429 the Venetian commander Silvestro Mocenigo attacked Gallipoli to destroy most of the Ottoman ships anchored there. And the Ottomans did not respond, limiting themselves to shows of force, sailing by the walls of Constantinople or by the Aegean islands. Pryor, *Geography, Technology, and War,* 175–76.

91. Archibald Lewis, "Islamic World and the Latin West," 836; Hess, "Evolution of the Ottoman Seaborne Empire," 1899–1901.

92. The numerical superiority of the Ottoman fleet concealed its technological inferiority to the navies of the Atlantic powers and even of the declining Venice. See Braudel, *Mediterranean and the Mediterranean World,* 1:445–47. For instance, after their 1571 defeat at Lepanto, where they lost two hundred out of three hundred ships, the Ottomans became, in Ranke's words, "conscious of the vices in their naval system." Ranke argues that those vices were in the social structure of the empire: the construction of ships was entrusted to slaves, who, careless in the choice of wood and the assembly of ships, produced unseaworthy vessels. Ranke, *Ottoman and the Spanish Empires,* 23. Ottoman ships continued to be inferior to the Venetian ones even in the seventeenth century. Furthermore, Ottoman galleys were unfit for navigation in the Atlantic and, most importantly, in the Indian Ocean, where the Portuguese navy controlled key sea lanes. For instance, a 1554 naval expedition launched from Egypt to defeat the Portuguese in the Indian Ocean and recapture the spice trade failed in part because the Ottoman fleet was not suited to the oceanic waters. See Pryor, *Geography, Technology, and War,* 180; and Cipolla, *Guns, Sails, and Empires,* 126–48.

93. Pears, "Ottoman Turks," 694; Inalcik, "Mehmed the Conqueror," 443.

94. For instance, a Venetian ship trying to bring grain to the besieged city of Constantinople was destroyed by Ottoman gunfire when it refused to stop by a castle to pay its dues. To emphasize his control over the straits, the sultan ordered the Venetian crew to be

beheaded and its captain impaled. Babinger, *Mehmed the Conqueror and His Time*, 77–79; Inalcik, "Mehmed the Conqueror," 422; Barbaro, *Diary of the Siege*, 10.

95. Pryor, *Geography, Technology, and War*, 177. Pryor also has written that the Ottoman maritime advantage was based on the "superior quality of the strategic bases acquired for the Ottoman navy and from the logistical advantages held by the Ottoman navy and corsairs in an age when maritime traffic continued to ply the age-old coastal trunk routes of the sea and when the main strike weapon at sea was still the oared galley" (192).

96. Ibid., 177. See also Babinger, *Mehmed the Conqueror and His Time*, 283–85.

97. In 1477, for example, an Ottoman contingent of ten thousand soldiers reached the region of Friuli, devastating the area. See Cremonesi, *La sfida turca*, 128–32.

98. See Shaw, *History of the Ottoman Empire*, 76.

99. Venice's overseas bases were easy to conquer because they were meant to be merely support ports for its fleet rather than centers of territorial dominion. In the sixteenth century "the frontier possessions of Venice were merely a string of tiny settlements in often archaic forts. The populations of the towns and the islands rarely exceeded a few thousand: in 1576, Zara had just over 7000 inhabitants, Spalato just under 4000.... The empire was insignificant, in demographic terms, compared with Venice and the Terraferma, the population of which was estimated about the same time as one and a half million." Braudel, *Mediterranean and the Mediterranean World*, 2:846.

100. Pryor, *Geography, Technology, and War*, 182.

101. The Spanish conquest of Grenada in 1492 pushed Muslims out of western Europe and into North Africa. A legacy of the Reconquista (Reconquest) was the creation of a large number of Muslim corsairs, many of whom had previously ruled southern Spain. Others, such as the Barbarossa brothers, were adventurers that established their bases on the thinly populated North African coast and spread terror in the western Mediterranean in the early sixteenth century. While the Ottoman sultan did not spur these corsairs, once they were established he approved of their actions and extended his benign protection over them. European powers tried to defeat them but failed several times. The most disastrous expedition was one against the Barbarossa in North Africa in 1518. The expedition, encouraged by the pope, ended in complete disaster. For a history of the Muslim corsairs in the sixteenth through eighteenth centuries see Currey, *Sea Wolves of the Mediterranean*.

102. For a brief history of Algiers as a center of piracy see Braudel, *Mediterranean and the Mediterranean World*, 2:884–87.

103. The activity of Muslim corsairs spread fear of a Turkish invasion of Europe. Indeed, the Ottoman navy was feared as far as away Poland and Russia. For example, in the second half of the sixteenth century rumors of an impending naval attack on Kiev (Ukraine) and Gdansk (Poland), however risible, were taken seriously. Both states had experienced several Ottoman land attacks and expected the Ottomans to be as strong on the sea as they were on land. See Abrahamowicz, "L'Europe orientale."

104. Braudel, *Mediterranean and the Mediterranean World*, 2:883.

105. For the relationship between the Ottoman state and the "bandits," or mercenaries, see also Barkey, *Bandits and Bureaucrats*, esp. 189-228.

106. Muslim corsairs, albeit the most famous and terrifying, were only some of the

many pirates in the Mediterranean during the sixteenth century. Christian pirates were just as daring and cruel. See Braudel, *Mediterranean and the Mediterranean World*, 2:870–80.

107. Cot, *Suleiman the Magnificent*, 107.

108. A few decades earlier, in 1480, Muhammed II the Conqueror had failed to take Rhodes and desisted from further attacks because at the time the island was relatively unimportant to the Ottoman Empire since it was not yet in control of Egypt. See ibid., 43–44. For the 1480 attack on Rhodes see Babinger, *Mehmed the Conqueror and His Time*, 396–400; and Hess, "Evolution of the Ottoman Seaborne Empire," 1904.

109. For this reason Cyprus was also seen by Venice as the last defense against the Ottomans. See Beeching, *Galleys at Lepanto*, 181; and Mazzarella, introduction to *La battaglia di Lepanto*, 14.

110. Lesure, *Lépante*, 25–31.

111. "The Spaniards, since the *Reconquista*, had taken control of most of the fortified positions on the North African coast to protect their communications with Sicily, which supplied them with grain, and to drive away the corsairs who raided their coasts." Cot, *Suleiman the Magnificent*, 102.

112. See Braudel, *Mediterranean and the Mediterranean World*, 2:987-92; and Canosa, *Lepanto*, 35–41.

113. Braudel, *Mediterranean and the Mediterranean World*, 2:1088.

114. In 1572, a year after Lepanto, it was reported that the sultan had already built 150 galleys, restoring the numerical grandeur of his fleet. Yet these galleys were poorly built and weakly armed, made as a show of force, "a high card in the diplomatic game." Beeching, *Galleys at Lepanto*, 228. Moreover, after the massacre at Lepanto the Ottoman navy had a chronic lack of manpower. The manning of the galleys became so difficult that simple oarsmen were elevated to the rank of officer. See Canosa, *Lepanto*, 277; and Lesure, *Lépante*, 230–32. For a different opinion see Cremonesi, *La sfida turca*, 309–10.

115. Braudel, *Mediterranean and the Mediterranean World*, 2:1088.

116. "Not only have the Turks lost that arrogant impression that Christians would not have the courage to face them in battle but now they [the Turks] are afraid to engage our forces, admitting that their galleys are considerably inferior to ours." Marco Antonio Barbaro, quoted in Ranke, *Ottoman and the Spanish Empires*, 23n, my translation.

117. Voltaire famously observed that if one looked at the 1573 peace, Venice appeared to have been the loser at Lepanto. However, the indemnity the Venetians agreed to pay the Ottomans was less than the cost of manning a fleet for a year. Moreover, Venice recovered all of its Aegean empire except Cyprus. The separate peace signed by Venice with the sultan ended the Holy League, leaving Spain the burden of fighting the Ottomans in the Mediterranean. See Beeching, *Galleys at Lepanto*, 231–32; and Faroqhi, "Venetian Presence in the Ottoman Empire."

118. See, e.g., Shaw, *History of the Ottoman Empire*, 178; and Canosa, *Lepanto*, 321–30. Also, though sharing no borders with the expanding Ottoman Empire and being an uneasy neighbor of the Spanish and Austrian powers, France had been even an ally of the Ottomans in the fifteenth century. In the sixteenth century it planned with the Ottomans a joint invasion of Italy (1537) and joined the sultan in an anti-Spanish alliance (1553). It was not a member of the Holy League. See Cremonesi, *La sfida turca*, 218–20.

119. Girolamo Diedo, letter of December 1571, in Caetani and Diedo, *La battaglia di Lepanto*, 224.

120. See Tamborra, "Dopo Lepanto."

121. See Beeching, *Galleys at Lepanto*, 158.

122. Two land routes linked the Red Sea and the Persian Gulf with the territories of the sultan. The first route started in Basra, on the Persian Gulf, and reached Anatolia through Baghdad (and later also Yerevan). The second connected ports on the Red Sea with Alexandria, Cairo, and other cities on the southeastern Mediterranean shore. The flow of goods on both routes was conditional on free sea lanes in the Persian Gulf and the Rea Sea and, farther away, on the Indian Ocean. Muslim merchants traveled on these seas, carrying Asian goods from India to ports on the Red Sea and the Persian Gulf. By diverting trade directly to Portugal via the Cape Horn, the Portuguese threatened to cut this maritime connection with India. See Shaw, *History of the Ottoman Empire*, 99; and Stripling, *Ottoman Turks and the Arabs*, 15. In 1507, for instance, no Indian goods arrived at Mamluk ports because the Red Sea had been closed by the Portuguese fleet. Ibid., 31–32.

123. The geopolitical change also diminished the interest of European merchants, especially from Genoa and Venice, in trading with the Ottomans. The expansionary policy of the Ottoman Empire in Europe and in the Mediterranean might have had some disruptive influence on trade but "can hardly explain the movement of the Genoese [and other Europeans] out of the Ottoman market. Presumably this was due to other factors unconnected with any Ottoman policy and resulting more from Genoese motivation to find richer pickins in the newly discovered markets of the New World." Fleet, *European and Islamic Trade*, 133.

124. Braudel, *Mediterranean and the Mediterranean World*, 2:844. "Turkey's swing to the East was balanced by Spain's swing to the West: gigantic shifts which the history of events cannot by its very nature explain. . . . [The] Hispanic bloc and the Ottoman bloc, so long locked together in a struggle for the Mediterranean, at last disengaged their forces and at a stroke, the inland sea was free from that international war which had from 1550–1580 been its major feature." Ibid., 1185.

125. From 1644 to 1669 the Ottomans engaged in a vicious war with Venice, conquering Crete. But this success did not increase Ottoman sea power. On the contrary, the conflict underscored the Ottomans' naval weaknesses and their inability to control the sea.

126. Cot, *Suleiman the Magnificent*, 90 and n. 15.

127. Quoted in Bernard Lewis, *Muslim Discovery of Europe*, 34.

128. Ranke, *Ottoman and the Spanish Empires*, 23n.

129. Palmer, *Decline and Fall of the Ottoman Empire*, 16; Goodwin, *Janissaries*, 91.

130. An indication of the importance of the Ottomans' desire to keep trade routes open rather than merely conquer territories and expand where there was less resistance was fact that the Ottomans helped the Mamluks in 1509 to build their own fleet in the Red Sea. The Ottomans gave the Mamluks timber to build fully equipped ships, three hundred guns, three thousand oars, and sails. But the inexperienced Mamluk navy was destroyed by the Portuguese fleet. Goodwin, *Janissaries*, 124; Stripling, *Ottoman Turks and the Arabs*, 32; Hess, "Evolution of the Ottoman Seaborne Empire," 1907–9.

131. Stripling, *Ottoman Turks and the Arabs*, 38–55.

132. It is important to note that the sultans were not motivated by religious concerns. The Ottomans did not expand in the Red Sea to defend the Muslim holy cities against Portuguese attacks. Had this been the case, the sultans would not have abandoned their naval operations in the Red Sea in favor of a conquest of Hungary in 1526. See also Hess, "Evolution of the Ottoman Seaborne Empire," 1917–18.

133. The Portuguese navy was unparalleled in terms of both logistics and skill. While the Ottomans had only four good ports in Arabia and had to rely on Anatolia and Europe for supplies for its fleet, the Portuguese controlled several ports and had close access to resources in the Indian Ocean. Moreover, whereas Ottoman sailors were experienced in Mediterranean seafaring, the Portuguese had developed superior skills in navigating the ocean waters around Africa and in Asia. The naval tactics of the Ottomans also revealed their land mentality. When Ottoman ships encountered Portuguese ships, they were arranged in a tight group, close to the shore, from where Ottoman cannons of the army or of some fort could support them. (According to some, the Ottoman expedition failed because of the incompetence of its commander, Suleiman Pasha, rather than because the Ottomans were militarily inferior to the Portuguese.) Stripling, *Ottoman Turks and the Arabs*, 92–96; see also Hess, "Evolution of the Ottoman Seaborne Empire," 1910.

134. Stripling writes: "Few enterprises costing so much as the fleet of 1537–1539 have had such a negligible result." Stripling, *Ottoman Turks and the Arabs*, 92. See also Cot, *Suleiman the Magnificent*, 194–95.

135. Ottoman sultans had previously devoted some efforts to stabilizing their southern frontier, especially with Persia. The Muslim sect of the Safavids, vehemently hostile to the Ottomans, controlled Baghdad and the surrounding region. Because they presented a constant threat to the Ottoman Empire, the Safavids often received assistance from European powers eager to divert the sultans' attention from their own territories. In fact, as mentioned above, on several occasions both the pope and the Portuguese sent envoys and military supplies, including cannons, to the Persian shahs in exchange for help against the Ottomans. Nonetheless, until the sixteenth century Ottoman efforts were limited to punitive raids. See Goodwin, *Janissaries*, 122; and Stripling, *Ottoman Turks and the Arabs*, 79.

136. See Stripling, *Ottoman Turks and the Arabs*, 78–79.

137. That an Ottoman dignitary, who would have been one the most well fed members of an Ottoman expedition, died of starvation during the march to Baghdad attests to the effectiveness of this "scorched earth policy." Cot, *Suleiman the Magnificent*, 91.

138. Porciatti, *Dall'impero ottomano alla nuova Turchia*, 57.

139. Shaw, *History of the Ottoman Empire*, 107.

140. Veinstein, "Commercial Relations," 96–97, 100.

141. "Traders far more than conquerors, the Kings of Portugal only snatched the coastal bases they needed for commerce. The sultans, on the other hand, occupied vast inhabited territories to extract fiscal and material resources. Their interests were so diverse and concerned such different areas that neither side ever felt it had to make the immense effort needed for a major campaign. Paying little attention to what happened in the Indian Ocean and on far-away islands, Suleiman began to lose interest in Asia." Cot, *Suleiman the Magnificent*, 198.

142. As described below, the Ottomans continued to push into Europe while decreasing

their efforts in the south. In 1521 the Ottomans conquered Belgrade, and in 1526 Suleiman the Magnificent invaded Hungary.

143. By controlling the shores of the Red Sea, the Ottomans succeeded in maintaining a flow of Asian goods. "It was only in the seventeenth century and after, when [the Portuguese] were replaced by the far stronger fleets of England and the Netherlands, that the old routes were finally closed and the Middle East was thrown into an economic depression from which it recovered only in modern times." Shaw, *History of the Ottoman Empire,* 107. See also Cot, *Suleiman the Magnificent,* 197; Stripling, *Ottoman Turks and the Arabs,* 102–7; and Cizakca, "Price History and the Bursa Silk Industry."

144. "The Ottomans were in fact less interested in maritime commerce than the conquest of fresh territories where they could extract revenues from taxes on the produce of the earth and the exchange of merchandise. Operations in the Balkans and Central Europe brought them greater profits than major naval expeditions into warm seas could have done." Cot, *Suleiman the Magnificent,* 192. See also Tamborra, "Dopo Lepanto," 373–74.

145. Lieven, *Empire,* 130.

146. Cot, *Suleiman the Magnificent,* 117.

147. This period also marks the largest naval expansion of Spain. As a result of Spanish maritime supremacy in the Mediterranean, other states, among them the papacy and Genoa, "virtually demilitarized themselves, choosing to dismantle their own fleet and to rely instead on Philip's galleys for their defense." Parker, *Grand Strategy of Philip II,* 84.

148. In the 1566 campaign in Hungary Suleiman the Magnificent died, leaving behind the largest empire in Europe. Constantinople had 700,000 inhabitants, compared with London's 120,000 and Seville's 100,000. See Cot, *Suleiman the Magnificent,* 299.

149. The stability of the central European front was also brought about by slow underlying changes in the military structure of European states. In fact, the realization that the Ottomans could not be annihilated through a crusade led to important changes in the European military organization and strategy. Since the fifteenth century the Ottomans had consolidated their possessions in the Balkans and been in constant contact with the Habsburg and even northern Italian lands. The sultans usually conducted a major operation for a few months and then retreated to the heart of their empire or to their southern frontier. But local Ottoman lords continued to organize marauding expeditions that penetrated deeply into the European territories—as far as Friuli and Vienna. The medieval way of waging war, through engagements by armies raised for a short period of time, was insufficient to deal with the continuous forays of the Ottomans. As a result, two important changes occurred on the European side. First, mercenary armies were no longer sufficient to cope with the constant threat of Ottoman raids. Maintaining a standing army, perennially alert, was the only way to defending the frontier regions. Renaissance military thinkers, such as Machiavelli, had foreseen such necessity, even though they were wrong on many other aspects of the art of war. Second, in order to build a stable frontier it was necessary to establish military colonies or fortifications similar to those built by the Romans along their northern frontier. Aventinus (1477–1534), a Bavarian historian, proposed establishing a belt of military colonies along the frontier with the Ottomans by giving land to peasants and to soldiers, thereby creating incentives for them to defend their lands from the Turkish forays. "Aristotle already had realized the necessity of giving the common man something

to fight for," he wrote. And the Holy Roman Empire should follow "the ancient Romans," who "always maintained a strong body of permanent soldiers, whom they endowed with land, and who, together with their families, were granted the same special and privileged status which the clergy enjoys in our day." Quoted in Rothenberg, "Aventinus," 64. See also Gilbert, "Machiavelli," 11–15; and McNeill, *Pursuit of Power,* 90–94.

150. See Pajewski, *Buńczuk i koncerz,* 62–85.

151. Ibid., 148–64.

152. Fuller, *Strategy and Power in Russia,* 15. Crimea and Ukraine were also less significant than the Danube valley because of their geographic features. From the south, "the approaches to the Crimea were guarded by 360 square miles that were, essentially, an inhospitable desert" (107). In the early 1700s a German general even used camels during his expedition in the region. Similarly, the Moldavian plains, south of Ukraine, were "transsected by numerous gorges and ravines" (107). Given these geographic obstacles, similar to those in Persia, the Ottomans had trouble expanding in this direction.

153. "A survey of the military events of the years 1683 to 1699 shows conclusively that the center of the fighting was located in Hungary. The Polish, Russian, and Venetian conflicts were peripheral. . . . Neither Poland, nor Russia, nor Venice represented so direct a threat to the Turkish Empire as did Austria . . . The territory that was menaced by the peripheral powers was not of prime importance, whereas Hungary, formerly the regular road to further Ottoman advance, was now the road for attacks that threatened to be fatal." Munson, *Last Crusade,* 120.

154. The geostrategic importance of Vienna had been well known since Roman times. Marcus Aurelius considered Vindobona (near modern Vienna) the point linking east and west, south and north. See Moczulski, *Geopolityka,* 174–75.

155. Moreover, the Danube valley was closer to Ottoman bases in Constantinople. Given the seasonal nature of wars, Ottoman sultans could not project power easily to Ukraine and Crimea in time to conduct lengthy campaigns. In the Danube a combination of roads and rivers allowed the Ottomans to send an army in the spring (even in midsummer) and still reach Vienna before winter set in. See also Palmer, *Decline and Fall of the Ottoman Empire,* 17.

156. See Munson, *Last Crusade,* 8.

157. Had the Ottomans succeeded in occupying Vienna and defeating the Habsburgs in Austria, France would have become the greatest power in Europe. France had been a tradition ally of the Ottomans, and even after a resurgence of a crusading spirit with Louis XIV, it maintained friendly relationship with the sultans, hoping to obtain commercial benefits and to balance the Habsburgs. In 1673 France signed a commercial agreement with Constantinople. See Barker, *Double Eagle and Crescent,* 67, 375–76.

158. Munson, *Last Crusade,* 7; Sobieski, *Jan Sobieski: Listy do Marysieńki,* 495–97; Wójcik, *Jan Sobieski,* 314–16.

159. In a letter to one of his commanders, the Polish king John Sobieski argued that Poland had to fight for Austria and Vienna because it was easier logistically and diplomatically. Logistically, it was easier because the Polish army could live off the wealthy Austrian lands without damaging their own. Diplomatically, it was easier because as a defender of Vienna, Poland was supported by other European states. Had the Ottomans

attacked Poland directly, no great power would have been on the Polish side, considering the danger to Vienna a much greater threat to Europe. Sobieski to Mikołaj Sieniawski, in Wójcik, *Jan Sobieski*, 322.

160. In Persia the logistical problems of imperial extension were very similar. Projecting power there was even more difficult because of the lack of well-maintained, easy routes. To cross a desert with an army of forty thousand was simply suicidal. "If they set out at the beginning of spring, the sultan's troops would not reach the shah's territories before June. . . . The campaign could not last more than three or four months—to travel back through snow always led to thousands of casualties." Cot, *Suleiman the Magnificent*, 301.

161. Barker, *Double Eagle and Crescent*, 18.

162. Russia joined the alliance only in 1686, after signing a treaty of "perpetual peace" with Poland, considered to be a more dangerous enemy than the Ottomans. Because of mutual animosities, the Holy League was a fragile alliance. Moreover, the most dangerous threat to the Ottomans came from Austria, in the Danube region, while the other theaters of actions were peripheral. For example, Russia's attack in Crimea (1687–89) was badly organized and had no significant strategic consequences. See Fuller, *Strategy and Power in Russia*, 17–21; and Anisimov, "Imperial Heritage of Peter the Great." On the fragility of the Holy Alliance, see Pajewski, *Buńczuk i koncerz*, 257.

163. For example, in 1688 the Ottoman army did not leave the capital until 29 June, and five years later not until 5 July, leaving only a few months to conduct military operations on the northern frontier.

164. Munson, *Last Crusade*, 143.

165. See also Bernard Lewis, *Muslim Discovery of Europe*, 42–43.

166. Paul Wittek, *Rise of the Ottoman Empire*, 3.

167. Ranke, *Ottoman and the Spanish Empires*, 24.

168. Spykman, "Geography and Foreign Policy, II," 224.

Six • *The Geostrategy of Ming China (1364–1644)*

1. Hucker, *Ming Dynasty*, 1; Dreyer, *Early Ming China*, 2; Farmer, *Early Ming Government*, 3–5. The Manchu dynasty, although based on a non-Sinitic tribe, adopted the institutions that it inherited from the Ming period. See Mancall, "Ch'ing Tribute System," 85. For a historical examination of China's foreign policy see also Swaine and Tellis, *Interpreting China's Grand Strategy*; and Kang, "Hierarchy in Asian International Relations."

2. The argument suggesting that China has a distinct (and peaceful) culture that informs its foreign policy is examined in greater detail below.

3. In fact, some argue that from Roman times China was stronger than any European state. The direction of trade, which privileged Chinese exports to Europe, seems to indicate the greater commercial development of China. "The greater demand for Chinese goods abroad than for foreign goods in China was to remain characteristic of China's trade until the nineteenth century, in large part because of superior Chinese technology as compared with other areas during most of this period." Fairbank, Reischauer, and Craig, *East Asia*, 76. Some argue that "during the late fifteenth century China was still the greatest economic power on earth. It had a population probably in excess of 100 million, a prodigiously

productive agricultural sector, a vast and sophisticated domestic trading network, and hand-icraft industries superior in just about every way to anything known in other parts of Eurasia." Atwell, "Ming China," 378. Of course, if we consider war the ultimate test of relative power, we cannot proffer a final judgment on China's superiority over Europe. See also Deng, "Critical Survey."

4. On dynasties in Chinese historiography see Mote, *Imperial China*, 563; and McNeill, "Changing Shape of World History," 9.

5. See McNeill, *Pursuit of Power*, 24–44; and Fairbank, Reischauer, and Craig, *East Asia*, 134–35.

6. Fairbank, Reischauer, and Craig, *East Asia*, 136. On the Song and Mongol navies see also McNeill, *Pursuit of Power*, 42–44.

7. Quoted in Hudson, *Europe and China*, 134.

8. Marco Polo (1245–1324) was in China about 1270; the Arab traveler Ibn Battuta (1304–77) also reached China through Central Asia. It is interesting that there is no trace of these European explorers in Chinese sources, while they were well known, albeit not always believed, in Europe. "The earliest accounts of China . . . aroused nothing but incredulity, so contrary were they to European preconceptions and so like fairy tales." Ibid., 162–63. For instance, travelers described cities three times as large as Venice and seaports that were the greatest in the world. See Fletcher, "China and Central Asia," 206; and Spence, *Chan's Great Continent*, 1–18.

9. Moreover, the Mongols also had to deal with their land borders. See also Lam, "Intervention versus Tribute in Sino-Vietnamese Relations," 167.

10. See Hucker, *Ming Dynasty*, 5.

11. Ibid.

12. Hudson, *Europe and China*, 154.

13. Farmer, *Early Ming Government*, 28–29.

14. For a more detailed examination of the competing rebellious groups see Mote, *Imperial China*, 520–33.

15. Dreyer, *Early Ming China*, 63. For the rise of the Ming see also ibid., 12–64.

16. See also Hucker, *Ming Dynasty*, 23–26; Sabattini and Santangelo, *Storia della Cina*, 474–75; Mote, *Imperial China*, 518n2; and Dardess, "Transformations of Messianic Revolt." I examine the "cultural distinctiveness" argument below.

17. The willingness to accept Mongol leaders and interethnic social mobility also explain why the Ming army achieved such great military success in the 1350s and 1360s. F. W. Mote observes that Chinese society had no class, such as the European nobility, that could transmit military knowledge from generation to generation. Nonetheless, Ming command-ers proved to be extraordinarily successful on the battlefield because they were allowed to rise through the social ladder on the basis of their skills, unimpeded by their ethnic or factional origin. See Mote, *Imperial China*, 555–57; and Serruys, "Mongols Ennobled during the Early Ming."

18. Hucker, *Ming Dynasty*, 79. See also Dreyer, *Early Ming China*, 155. It is interesting that after the conquest of the Mongol capital Tatu, Ming commanders sealed and stored all the archives of the defeated dynasty and protected all the buildings. See Mote, *Imperial China*, 563.

19. McNeill, *Pursuit of Power*, 44.

20. The attention paid to the Mongols, however, was detrimental to China's relations with Central Asian tribes, decreasing its contacts with and influence over this strategic region connecting East Asia with Europe by land.

21. In part the rebellions throughout China (especially in the south) were caused by the strengthening of the central administration. See Dreyer, *Early Ming China*, 110–11.

22. Ibid., 74–76.

23. Waldron, "Problem of the Great Wall." Waldron describes the history and the historiography of the Wall of China. The rise of the wall—or walls—shows that the threat from the northern frontiers preceded the rise of the Ming dynasty. Most importantly, it also shows that the walls were ineffective against a mobile enemy and never stopped an invasion from the north. The Mongols, for instance, "encountered no Great Wall when they conquered China." Ibid., 656.

24. Dreyer, *Early Ming China*, 143–48.

25. Ibid., 170.

26. Ibid., 171–72.

27. Suzuki, "China's Relations with Inner Asia," 192.

28. In Mote's assessment, "Although it occasionally sponsored Chinese settlement beyond the Wall, the Chinese state never had hopes of doing much more than extending its influence defensively. It did not conceive of Chinese power as extending to direct administration of territories lying much beyond the Wall, if at all; neither actually nor ideally were even the nearer steppe territories to be integral parts of the Chinese state. Nor were they to be colonies, to be conquered and maintained by force of arms, and assigned revenue quotas essential to the state, like those the Romans established along their northern frontier." Mote, "T'u-mu Incident of 1449," 245.

29. In addition, because of its need to concentrate on the northern frontier Peking failed to maintain a link with the land routes in Central Asia. During the reigns of Hung-Wu and Yung-lo relations with Central Asia worsened. The Chinese rulers assumed that Central Asian tribes were under the authority of their court. Often the results of such assumptions were almost catastrophic. For instance, as mentioned earlier, Tamerlane, insulted by the Ming ruler and apparently concerned with reports of Chinese persecution of Muslims, organized an expedition against China in 1404. Fortunately for China, he died suddenly. But even after Yung-lo's death China's military decline led to a gradual worsening of relations with Central Asia. Envoys and merchants from Central Asia began to demand higher prices for their goods while refusing Ming money in lieu of expensive goods. See Rossabi, "Ming and Inner Asia," 246–50.

30. Lattimore, *Inner Asian Frontiers of China*, 239, 241, 470, 240.

31. Farmer, *Early Ming Government*, 43.

32. Fairbank, Reischauer, and Craig, *East Asia*, 201.

33. The focus on the north also altered the nature of Chinese political preoccupations. The north was "where the government was more concerned with frontier control than with resource control, where considerations of security and defense outweighed those of civil administration and tax collection." Farmer, *Early Ming Government*, 10.

34. The Grand Canal was the lifeline of Peking, and its importance increased when sea

transport was abandoned. In 1487 a Ming official warned his superiors of the dangers of relying exclusively on the canal: "The nation's capital is now at Peking, that is to say at the extreme north, while the inflow of taxes comes entirely from the south-east. The Hui-t'ung Canal may be likened to a man's throat. If food cannot be swallowed for a single day, death ensues at once. . . . I would request that a sea route be opened, following the old Yuan dynasty route, and that it be operated in parallel with the transport of inland waterway in the autumn season when agricultural work is slack. . . . This is a plan to forestall future disaster." Quoted in Elvin, *Pattern of the Chinese Past*, 105. See also Dreyer, *Early Ming China*, 184.

35. The domestic confusion caused by Hung-Wu's death emboldened the Mongols. In the first years of the fifteenth century the Mongol leader, Tamerlane (Timur), desisted abruptly from his victorious campaign in Europe and the Middle East (in which he defeated even the rising power of the Ottomans [see chapter 5]) to devote all of his energies to the conquest of China. It is still unclear what his motivations were regarding China, but in part he was irritated by the several requests of tribute from the Ming court. In 1405, however, Tamerlane died, and the preparations for the invasion of China were abandoned. What might have happened is one of the greatest "what if's" of history.

36. For instance, during the 1421 campaign Yung-lo led an expedition to punish the Mongol leader Arughtai. As in other campaigns (reminiscent also of the Ottoman ones against the Safavids in Persia at roughly at the same time), the Chinese army successfully engaged Mongol detachments in several battles but did not succeed in meeting the main force of the Mongols. Arughtai chose not to encounter the Chinese, retreating to Outer Mongolia, where Yung-lo did not want to venture. Although in the final analysis the Ming army failed to obtain a decisive victory, the emperor returned to Peking claiming victory. During the winter Arughtai returned from his self-imposed exile in the middle of the steppes, reconstituted his hordes, and renewed his attacks on China. In 1424 the Chinese army had to return to Mongolia in search of Arughtai's hordes but again failed to find and destroy him. While returning to Peking, Yung-lo died outside the Great Wall. See Dreyer, *Early Ming China*, 179–82.

37. Waldron, "Chinese Strategy," 105.

38. Dreyer, *Early Ming China*, 182.

39. Waldron, "Chinese Strategy," 99.

40. See, e.g., Mote, "T'u-mu Incident of 1449," 271–72; and idem, *Imperial China*, 692–93.

41. Rossabi, "Ming and Inner Asia," 261.

42. It is, of course, doubtful whether unobstructed trade would have extirpated the Mongol threat. First, as observed, Chinese policy toward the Jurchens in the long term created a power next to China. In the sixteenth century the Jurchens became a threat to Peking, and in the mid-seventeenth century they conquered China and founded the Manchu dynasty. Second, the Jurchens were fundamentally different from the Mongols because they led a sedentary life, based on agriculture, as opposed to the nomadic lifestyle of the Mongolian tribes. According to some, notably Owen Lattimore, the relations between agricultural China and the nomadic Mongols were more fissiparous than those between China and the Jurchens because of the different economies of China and Mongolia. In other

words, a conflict between an agricultural (China) and a nomadic society (Mongols) was more difficult to reconcile than one between two agricultural societies (China, Jurchens).

43. Mote, "T'u-mu Incident of 1449," 245.

44. Rossabi, "Ming and Inner Asia," 231.

45. The power of the Ming court was such that it could afford to sacrifice the imprisoned emperor to the Mongols. When Esen arrived at the gates of Peking, he was told that the Ming emperor he was holding hostage was of no value to the Chinese. "It is the altars of the Earth and of Grain [i.e., the dynasty and nation] that are of great importance, while the ruler [as an individual] is unimportant." Quoted in Mote, "T'u-mu Incident of 1449," 266.

46. Edward Dreyer argues that Tu-mu was only the last in a series of events that impacted the course of Ming foreign policy. The withdrawal from Vietnam in 1427 marked the beginning of a process of change that resulted in an increasing self-imposed isolation of China. While fundamentally I do not disagree with Dreyer, it seems quite clearly that Tu-mu was a more scarring event because it involved the slaughter of thousands of Chinese soldiers, the imprisonment of the emperor, and the siege of Peking and above all because it was all done at the hands of the Mongols, who were regarded as China's most threatening enemy. As Arthur Waldron points out, the Tu-Mu battle marked "the last time a Chinese army actually left China proper and headed into the steppe to deal with the nomads." Waldron, "Chinese Strategy," 105. See also Dreyer, *Early Ming China*, 230; Mote, "T'u-Mu Incident of 1449," 267–72; and Lorge, "War and Warfare in China."

47. Part of the buffer zone had been abandoned during Yong-lo's reign as payment of a political debt to the Urianghai Mongols for supporting the emperor during his domestic struggle for power. But most of the forward defense system was abandoned only after Tu-Mu.

48. Quoted in Mote, "T'u-Mu Incident of 1449," 270.

49. Yet, according to Mote, "the northern border remained the source of an immense psychological threat, the major focus of a defense-minded government, and an unsolved problem. . . ." Although it was needed, "it turned the state away from either a decisive military solution or reasonable nonmilitary alternatives." Mote, "T'u-Mu Incident of 1449," 268–69.

50. Elvin, *Pattern of the Chinese Past*, 101.

51. It is true that, as Waldron observes, the Great Wall was not the product of a single strategic decision. "There had been no day when the Grand Secretaries said, 'let us build a Great Wall.'" Waldron, "Problem of the Great Wall," 660. It is not part of a coherently planned defensive structure but a series of walls built over several centuries. The most impressive part of the wall, near Peking, was erected during the late Ming period, after the Tu-Mu disaster. Before then the fortifications were part of a more balanced strategy that included outposts and expeditions in the steppes. see also ibid., 643–63.

52. Ibid., 661.

53. Mote also observes that today, paradoxically, the Great Wall is considered a symbol of China's greatness, not of a strategic failure of historical proportions. See Mote, *Imperial China*, 693.

54. Kenneth R. Hall, "Economic History of Early Southeast Asia," 261.

55. In Hung-Wu's words: "I am concerned that future generations might abuse China's wealth and power and covet the military glories of the moment to send armies into the field without reason and cause a loss of life. May they be sharply reminded that this is forbidden. As for the *hu* and *jung* barbarians who threaten China in the north and west, they are always a danger along our frontiers. Good generals must be picked and soldiers trained to prepare carefully against them." Quoted in Gungwu, "Ming Foreign Relations," 311–12.

56. See Dreyer, *Early Ming China*, 206.

57. According to Wang Gungwu, "The first Ming emperor believed in tight centralized control over all matters pertaining to relations beyond the borders of his empire. . . . He was anxious to control all foreign trade so as to ensure that trading along sensitive frontiers would not disturb the law and order of his realm." Gungwu, "Ming Foreign Relations," 307. See also Farmer, *Early Ming Government*, 11.

58. Gungwu, "Ming Foreign Relations," 302.

59. Dreyer, *Early Ming China*, 209.

60. See also Mote, *Imperial China*, 612–13.

61. Dreyer, *Early Ming China*, 230.

62. See Levathes, *When China Ruled the Seas*, 42–44; and Filesi, *China and Africa*, 5.

63. For a history of Song and Mongol naval power see Lo, "Emergence of China as a Sea Power."

64. See Levathes, *When China Ruled the Seas*, 50–54.

65. Elvin, *Pattern of the Chinese Past*, 220; Kenneth R. Hall, "Economic History of Early Southeast Asia," 217.

66. Hudson, *Europe and China*, 164.

67. Elvin, *Pattern of the Chinese Past*, 217.

68. Fairbank has argued the exact opposite, that is, that Confucian beliefs were behind Ming interests in maritime trade. Ming attempts to develop a trade system in Southeast Asia where motivating by the desire to establish a tribute system that acknowledge the cultural superiority of China. Thus, the expeditions led by the eunuch Cheng Ho (see below) were "an effort to bring the sources of Chinese maritime trade back into the formal structure of the tributary system so as to make the facts of foreign trade square with the theory that all places in contact with China were tributary to her." Fairbank, "Tributary Trade," 143.

69. Elvin, *Pattern of the Chinese Past*, 217.

70. Ibid., 231; Filesi, *China and Africa*, 33.

71. Sakai, "Ryuku (Liu-Ch'iu) Islands," 126–27.

72. Filesi, *China and Africa*, 33.

73. Elvin, *Pattern of the Chinese Past*, 220. See also Filesi, *China and Africa*, 52–55; and Parry, *Discovery of the Sea*.

74. Levathes, *When China Ruled the Seas*, 88; Sakai, "Ryuku (Liu-Ch'iu) Islands," 128.

75. Levathes, *When China Ruled the Seas*, 140.

76. For a description of sea routes and how they connected with internal routes in China see Reischauer, "Notes on T'ang Dynasty Sea Routes."

77. Levathes, *When China Ruled the Seas*, 88–98; Kenneth R. Hall, "Economic History of Early Southeast Asia," 227.

78. Levathes, *When China Ruled the Seas,* 109–10; Dreyer, *Early Ming China,* 199–200.

79. Levathes, *When China Ruled the Seas,* 142.

80. Elvin, *Pattern of the Chinese Past,* 220.

81. Johnston, *Cultural Realism,* ix. See also Burles and Shulsky, *Patterns in China's Use of Force,* 79–93.

82. Johnston, *Cultural Realism,* 61.

83. As Confucius said, "Govern the people by regulations, keep order among them by chastisements, and they will flee from you, and lose all self-respect. Govern them by moral force, keep order among them by ritual and they will keep their self-respect and come to you of their own accord." Quoted in Hucker, *Traditional Chinese State in Ming Times,* 76. An underlying assumption of the argument that Chinese international behavior was shaped by a Confucian culture is that Chinese culture remained static, with predominantly Confucian traits. Such an assumption is overly simplistic, especially in light of the interactions between China and Europeans from 1500 on. See also Mungello, *Great Encounter.* Fairbank is probably the most well known historian of China to have advocated this "cultural" explanation. For a critique see Johnston, *Cultural Realism,* 253.

84. Mote, *Imperial China,* 616.

85. See Hucker, *Traditional Chinese State in Ming Times,* 61, 71. According to Hucker, the Ming state "was Confucian in the sense that it espoused Confucianism as its official ideology" (61). Confucianism translated into a defensive and peaceful foreign policy that was fundamentally different from that of European powers. This belief in a Chinese culture of peace was introduced to Europe by the merchants and missionaries that lived in China in the sixteenth and seventeenth centuries. For instance, the Jesuit Matteo Ricci wrote: "It seems to be quite remarkable when we stop to consider it, that in a kingdom of almost limitless expanse and innumerable population, and abounding in copious supplies of every description, though they have a well-equipped army and navy that could easily conquer the neighboring nations, neither the King nor his people ever think of waging a war of aggression. They are quite content with what they have and are not ambitious of conquest. In this respect they are much different from the people of Europe." Spence, *Memory Palace of Matteo Ricci,* 54–55. See also Hucker, *Traditional Chinese State in Ming Times,* 71. Ricci was also perplexed by the highly developed technical skills of the Chinese, which were used not for military purposes but for pyrotechnic shows. The Chinese "created fantastic shows of flowers and fruit and battles all spinning through the air, every one made out of these same fireworks; and one year when I was in Nanjing [1599] I estimated that in the month-long New Year celebrations they used up more saltpetre and gunpowder than we would need for a war lasting two or three years." Quoted in Spence, *Memory Palace of Matteo Ricci,* 45–46. For a brief overview of the various arguments explaining the decline of China see Eric Jones, "Real Question about China."

86. Fairbank, Reischauer, and Craig, *East Asia,* 178. See also Pelissier, *Awakening of China,* 12.

87. The cultural argument can also be used to justify the exact opposite, that is, the maritime expansion of China. For instance, Fairbank argues that Cheng Ho's expeditions were motivated by "the desire of the Ming court to perfect its claim to rule all men by bringing the maritime trading nations of the world into the traditional suzerain-vassal

relationship which was demanded by Confucian theory as an alternative to the direct rule of the Son of Heaven. Ming suzerainty was vigorously asserted over the tribes of Central Asia in the early fifteenth century. The Cheng Ho expeditions seem to have been equally an effort to assert Chinese suzerainty over rulers accessible by sea." Sakai, "Ryuku (Liu-Ch'iu) Islands," 128–29.

88. Mote argues that Cheng Ho's expeditions were the only state-sponsored attempts to create a sea empire, while all other maritime activities were undertaken mostly by private subjects. This is because "the Chinese state and its official policy relegated the seafaring activities of China's coastal population to a third-rate place in the nation's life." Hence, the end of Cheng Ho's expeditions was not a surprise but a return to the usual official policy of indifference toward the sea. However, the distinction between state-sponsored and private expeditions should not be exaggerated. First, China was a maritime power not only because of private interests. Both the Song and the Mongols were actively involved in the building and administration of the navy. The Ming emperors continued this policy, especially with Yung-lo. Second, Chinese merchants had roamed the South China Sea and the Indian Ocean for decades and centuries before Cheng Ho, and their activity was economically and politically beneficial to the Chinese empire. China had access, even though in private hands, to the major sea lanes of Asia. The state sponsored the expeditions in the early fifteenth century because those merchants either were going against China's interests (e.g., Chinese pirates in Malacca) or could not maintain control over the sea lanes. In the absence of a powerful navy, the sea lanes gradually fell into the hands of pirates and later, in the early sixteenth century, into the hands of European powers, cutting China off. See Mote, *Imperial China,* 717-19. On the government-sponsored maritime activities under Song, Mongol, and Ming rule see Lo, "Emergence of China as a Sea Power."

89. Johnston, *Cultural Realism,* 176.

90. McNeill, *Pursuit of Power,* 46. McNeill also argues that "land frontiers against steppe raiders and potential conquerors remained more important than anything happening in the ports where European ships put in." By the time the Chinese defeated the last steppe power in the mid-eighteenth century Europe was already firmly in control of the sea lanes and ports of Asia. See idem, "World History," 230–31.

91. This situation also created a new "macroregion" in the north that competed with and drained the traditional center of resources in the Yangtze basin. See also Skinner, "Presidential Address," 275–76, 284.

92. Levathes, *When China Ruled the Seas,* 144; Dreyer, *Early Ming China,* 184–85. See also Mote, *Imperial China,* 646–53.

93. This raises an interesting historical "what if." Dreyer suggests that had the imperial capital been in Nanjing rather than in Peking, Ming rulers might have been more interested in the maritime expeditions and the invasion of Vietnam. See Dreyer, *Early Ming China,* 202, 259. Similarly, as McNeill writes, "The overseas empire China had created by the early fifteenth century impels a westerner to think of what might have been if the Chinese had chosen to push their explorations still further. A Chinese Columbus might well have discovered the west coast of America half a century before the real Columbus blundered into Hispaniola in his vain search for Cathay. Assuredly, Chinese ships were seaworthy enough to sail across the Pacific and back. Indeed, if the likes of Cheng Ho's

expeditions had been renewed, Chinese navigators might well have rounded Africa and discovered Europe before Prince Henry the Navigator died (1460)." McNeill, *Pursuit of Power*, 45.

94. Dreyer, *Early Ming China*, 202.

95. As a result of Chinese withdrawal from the South China Sea and the Indian Ocean, regional trade declined, becoming more localized. Instead of an Asian market, there were trading regions in the western Indian Ocean, the Bay of Bengal, and the South China Sea. See Prakash, "Portuguese and the Dutch," 176.

96. Rossabi, "Ming and Inner Asia," 271.

97. Roth, "Manchu-Chinese Relationship," 10.

98. See So, *Japanese Piracy in Ming China*, 5–6; and Hucker, "Hu Tsung-hsien's Campaign," 276.

99. See So, *Japanese Piracy in Ming China*, 134–35.

100. Dreyer, *Early Ming China*, 233.

101. Hucker, "Hu Tsung-hsien's Campaign," 275.

102. Wills, "Maritime China," 215.

103. So, *Japanese Piracy in Ming China*, 154.

104. For a description of the conquest of Malacca see Parker, *Success Is Never Final*, 202–3.

105. See Parry, *Establishment of the European Hegemony*, 41–42; and Boxer, *Portuguese Seaborne Empire*, 46–48.

106. For a detailed description of Portugal's advance in East Asia see Wills, "Relations with Maritime Europeans," 335–41.

107. C. R. Boxer, for instance, argues that the Portuguese were admitted to the China trade on terms imposed by Peking, not vice versa. While formally true, China simply could not avoid the Portuguese influence in the region. Moreover, it benefited from Portugal's intermediary role in regional trade. See Boxer, *Portuguese Seaborne Empire*, 49.

108. Macao also had no farmland and was completely dependent on Chinese supplies of foodstuffs. See Wills, "Relations with Maritime Europeans," 346.

109. See Wills, "Maritime Asia," 102.

110. Cooper, "Mechanics of the Macao-Nagasaki Silk Trade," esp. 423-24. See also Bernard, "Les debuts des relations diplomatiques"; and Wills, "Relations with Maritime Europeans," 341.

111. Parry, "Transport and Trade Routes," 193. See also Boxer, *Portuguese Seaborne Empire*, 61.

112. For the difficulties of commercial relations between mainland China and Taiwan, as well as the extent of those relations, see Laurence G. Thompson, "Junk Passage across the Taiwan Strait."

113. See Wills, "Relations with Maritime Europeans," 369–75. Also, the diplomatic and commercial arrangements between Europeans and Chinese occurred mostly outside official Ming channels. "The evolution of such complex Sino-European accommodations as Macao, Manila, and the early network of missionaries and converts, owes a great deal to maritime Chinese both on the China coast and in foreign ports, to astute and realistic officials, and to statesmen and intellectuals who were much more open to novelty and to

interaction with foreigners than some clichés about Chinese culture would have us believe." Ibid., 375.

114. Boxer, *Portuguese Seaborne Empire*, 57–58.

115. Cipolla, *Guns, Sails, and Empires*, 123–24.

116. In 1624 a Chinese military treatise stated that "fire arms can be used on large boats by the waves make aiming very hard. Chances of hitting the enemy are very slender. Even if one enemy boat should be hit, the enemy would not thus incur severe losses. The purpose of having fire arms aboard is purely psychological, namely that of disheartening the enemy." Ibid., 125–26.

117. Wills, "Relations with Maritime Europeans," 335.

118. Wills, "Maritime China," 221.

119. Goldstone, "East and West in the Seventeenth Century," 115.

120. In the late sixteenth century few advocated an invasion and territorial conquest of China or other Asian countries. Most who did were Jesuit missionaries who had proselytizing goals and did not possess a keen sense of the geostrategic needs of Portugal or Spain. They were missionaries, not strategists, and their proposals were rarely taken seriously by the authorities, even though they tried to lure the Spanish emperor with promises of enormous wealth. As the Portuguese bishop of Mallacca argued in 1584, Spain should conquer, with a very limited expenditure of manpower, the city of Canton, which was "so rich and sumptuous, and all those other regions of the south [of China], which are many, and very great and very wealthy. And thus His Majesty will be the greatest lord that ever was in the world." Parker, *Success Is Never Final*, 27. See also Boxer, *Portuguese Conquest and Commerce*, 131–36.

This, however, does not that the European empires were not interested in probing China's military capabilities. In the sixteenth century missionaries reported their observations on military issues. Matteo Ricci, probably mirroring similar descriptions of other observers, noted that Chinese men were effeminate and incapable of fighting. "Rarely do they wound or kill each other, and even if they wanted to they don't have the means, because not only are there few soldiers, but most of them don't even have a knife in the house. In short one has no more to fear from them than one would have from any large crowd of men; and though in truth they have plenty of fortresses, and their cities are all walled against the attacks of thieves, they are not walls built along geometrical principles, and they have neither traverses nor ditches." Spence, *Memory Palace of Matteo Ricci*, 43–44. Such detailed descriptions of China's defenses were part of an ongoing debate on the benefits and possibilities of invading the mainland, but they never materialized into a strategic project.

121. Boxer, *Portuguese Seaborne Empire*, 52.

122. For instance, Parry writes that in 1569, when Goa was attacked from land, only "command of the sea—the only route by which reinforcements might come—saved the Portuguese garrisons from annihilation." The example of this event can be generalized to the history of the Portuguese seaborne empire in Asia. See Parry, *Establishment of the European Hegemony*, 91; and Boxer, *Portuguese Seaborne Empire*, 53.

123. Moreover, the profits derived from controlling trade financed Portugal's imperial

machine, which in turn maintained the command over the sea and commerce. The sym-biotic relationship between trade and military capability is well summarized by the words of a Dutchman, Jan Pietersz Coen: "You . . . should well know from experience that in Asia trade must be driven and maintained under the protection and favour of your own weap-ons, and that the weapons must be wielded from the profits gained by the trade; so that trade cannot be maintained without war, nor war without trade." Quoted in Boxer, *Por-tuguese Conquest and Commerce*, 3.

124. As their monopoly over regional trade increased, European powers' capacity to dictate prices increased correspondingly. For instance, in the first decades of the seven-teenth century, "once the Dutch monopoly was established, prices to the growers were fixed at minimum levels, and all Asian intermediary traders and ports were eliminated so that profits flowed only to the VOC [the United East India Company–Dutch]." Reid, "Seventeenth-Century Crisis," 648. See also Parry, *Establishment of the European Hegemony*, 161–62. Such policies, however, also had negative repercussions on the overall level of trade. To sustain themselves, local populations were forced to grow basic foodstuffs instead of the minimally priced spices or silk, decreasing the quantity of the most profitable goods.

125. "Asian shipping was allowed to ply as before, provided that a Portuguese licence . . . was taken out on payment by the ship owner or merchants concerned, and provided that spices and other designated goods paid customs dues at Goa, Ormuz or Malacca. Un-licensed ships in the Indian Ocean were liable to be seized or sunk if they met Portuguese ships, particularly if the former belonged to Muslim traders." Boxer, *Portuguese Seaborne Empire*, 48.

126. Cipolla, *Guns, Sails, and Empires*, 143. The revenues from these permits sustained the Portuguese empire in its conflict with the Dutch. "During the first decades of the seventeenth century the Portuguese commercial network in Asia, facing major changes in trade patterns as other European companies entered Asian waters, was sustainable by its revenues from selling licences and from the customs paid at Ormuz, Goa, Diu, Cochin, and Malacca." John W. Wittek, "Review Article," 868.

127. See Reid, "Seventeenth-Century Crisis," 646–49.

128. The importance of American silver for Ming China was enormous. Because of the enormous profits derived from this trade, Chinese merchants had a small colony in Manila, which they maintained despite its being repeatedly subject to massacres (e.g., in 1603 and 1639). Moreover, there was a tiny Chinese colony as far away as Mexico, probably brought there by Spanish galleons trading Chinese silk for American silver. See Dubbs and Smith, "Chinese in Mexico City."

129. Atwell, "Seventeenth-Century 'General Crisis' in East Asia?" 677.

130. There is a lively debate on what is called the "seventeenth-century crisis." The debate is far from being resolved, and everything, from the root and the extent to the very existence of the crisis, continues to be questioned and discussed. While I do not take a position on the merits of the debate and of the concept of a "Eurasian crisis," evidence seems to indicate that China declined not just because of the mid-seventeenth-century slump in the silver trade from Mexico. Indeed, I argue that its decline began much earlier, when Portuguese and, later, Spanish and Dutch ships started to take control of the trade

routes in Asia. For an overview of the debate see Goldstone, "East and West in the Seventeenth Century"; Reid, "Seventeenth-Century Crisis"; Atwell, "Some Observations on the 'Seventeenth-Century Crisis' "; idem, "Seventeenth-Century 'General Crisis' in East Asia?"; Richards, "Seventeenth-Century Crisis in South Asia"; and Steensgaard, "Seventeenth-Century Crisis."

131. See Atwell, "Ming China," 410; and Parry, *Establishment of the European Hegemony,* 153–56.

132. There is some truth to Mahan's argument that land powers cannot become sea powers, but land powers can become sea powers when they pursue a successful continental policy that stabilizes their borders. This is where Ming China's failure lay: China failed as a sea power not because it had long land borders but because it failed to secure them.

133. Wills, "Maritime China," 208.

134. In fact, China's shores were quite well suited to the pursuit of maritime activities. As a Jung-Pang Lo observed, "The long coastline of China, large sections of which, as in Shantung, Chekiang, Fukien and Kwangtung, are endowed with harbors and timber-clad mountains, and the adjoining seas enclosed by an island fringe, provide the physiographical conditions which favor and promote maritime activities. But the territorial vastness of China and her cohesion with the continent of Eurasia exercise so profound an influence that for long centuries the attention of the Chinese was occupied with internal problems and the defense of their land frontier on the north and northwest, the directions from which danger had historically threatened." Lo, "Emergence of China as a Sea Power," 495.

135. The Ottoman Empire was similarly handicapped by extensive land borders. But unlike China, the Ottomans pursued a mostly continental expansion because the most important centers of resources and trade routes were across its land borders. A continental expansion would have brought China no tangible geostrategic gains.

Seven • *Lessons for the United States*

1. The argument that war is becoming more costly and does not lead to clear strategic gains has deep roots. Perhaps its most famous enunciation was a book by Norman Angell, *The Great Illusion,* which had the misfortune of being published a few years before World War I. Other exponents of this argument are Rosecrance, *Rise of the Trading State;* and Mueller, *Retreat from Doomsday.* See also Kaysen, "Is War Obsolete?"; Liberman, "Spoils of Conquest"; Orme, "Utility of Force"; and Brooks, "Globalization of Production."

2. See, e.g., Paul, "States, Security Function, and the New Global Forces."

3. In 2000, 37 percent of all U.S. international trade was waterborne. See Department of Transportation, Bureau of Transportation Statistics, *Transportation Statistics Annual Report 2001,* 215. For another excellent, albeit dated source of data on maritime trade see U.S. Congress, Office of Technology Assessment, *Assessment of Maritime Technology and Trade.*

4. Posen, "Command of the Commons," 8.

5. Niebuhr, *Structure of Nations and Empires,* 299.

6. See also Department of Defense, *Quadrennial Defense Review Report,* 3–5.

7. National Intelligence Council, *Mapping the Future,* 47.

8. See also Fry, "End of the Continental Century."

9. See, e.g., Andrews-Speed, Liao, and Dannreuther, "Strategic Implications of China's Energy Needs."

10. See, e.g., Barron, "China's Strategic Modernization"; Mark A. Stokes, *China's Strategic Modernization;* and U.S. Department of Defense, *Annual Report on the Military Power of the People's Republic of China.*

11. Shambaugh, "China Engages Asia," 66.

12. Nor does China's other historical enemy, Japan, present a threat since it is under the American security umbrella. See Simon, "Is There a U.S. Strategy for East Asia?" 329.

13. On the differences between the geographic features of Europe and East Asia see Friedberg, "Will Europe's Past Be Asia's Future?" 154–55.

14. Arguably, the United States is better prepared for a future maritime challenge than for the past continental one because of its history. As Owen Lattimore observed, "American tradition emphasizes the importance of sea power, especially the great age of sea power which began with Columbus and led first to the expansion of Spain and Portugal and then to the empire building of England, Holland, and France, and the development of North America. . . . Even our continental history . . . was initiated by the crossing of the Atlantic; and from then on even our period of most active continental expansion was never free of the influences and effects of sea power, sea-borne commerce, the investment of European capital, and acceleration of population growth by the immigration of Europeans. We are less familiar with the modes of history in areas of vast expanse, with considerable populations, which are not merely 'continental' but continent-bound. The influence of sea power was not absent from the earlier history of China, India, and Persia, but it was not decisive until the coming of the Europeans. In Russian history, access to the sea was of early importance; but control over sea routes was an ambition of late development. The earlier history of Eurasia was continent-bound. Major routes of migration and trade led from one land to another without crossing salt water." Lattimore, "Inner Asian Frontiers," 24.

15. Along the same lines, the ancient historian Josephus observed that the Romans "hold their wide-flung empire as the prize of valour, not the gift of fortune." Rome (as well as all the other great powers) was built and maintained by the actions of its citizens and leaders, not by some abstract, external force, whether fate or geopolitics. Josephus, *Jewish War*, 194.

16. The United States is also dependent on the oceans for its power-projection capabilities. As a 2003 RAND study points out, "Virtually any port in the world can be reached by sea from the continental United States within 17 days at 25 knots. This is about the same amount of time as would be required to deploy one interim brigade combat team if roughly one-quarter of the strategic airlift fleet were devoted solely to its transport. Moreover, faster transport ships, with speeds up to 40 knots, which would reduce sailing time by more than a third, may enter the fleet by the end of this decade." Cliff and Shapiro, "Shift to Asia," 97.

17. See Mahan, *Interest of America in Sea Power.*

18. See U.S. Department of Transportation, Bureau of Transportation Statistics, *Transportation Statistics Annual Report 2001,* chap. 7.

19. Ibid., fig. 2.

20. One scenario that appears to be increasingly likely foresees an escalation of terror-

ist threats to maritime routes, especially in Indonesia and the Philippines. See, e.g., Banlaoi, "Maritime Terrorism in Southeast Asia." For an overview of maritime security issues in Asia see Valencia, *Proliferation Security Initiative*, 9–24.

21. Friedberg, "11 September," 41.

22. See Fukuyama, "Re-Envisioning Asia"; and Friedberg, "Ripe for Rivalry."

23. See, e.g., O'Malley, "Central Asia and South Caucasus."

24. For the argument that China can pose a serious threat without matching U.S. capabilities see Christensen, "Posing Problems without Catching Up."

25. See, e.g., Krepinevich, testimony.

26. Goldstein and Murray, "Undersea Dragons."

27. St. Augustine, *City of God*, 4.3.111.

Abernethy, David B. *The Dynamics of Global Dominance*. New Haven, CT: Yale University Press, 2000.

Abrahamowicz, Zygmunt. "L'Europe orientale et les états islamiques au temps de la bataille de Lépante." In *Il Mediterraneo nella seconda metà del '500 alla luce di Lepanto*, edited by Gino Benzoni, 19–31. Florence: Leo S. Olschki Editore, 1974.

Abu-Lughod, Janet L. *Before European Hegemony*. New York: Oxford University Press, 1989.

Acemoglu, Daron, Simon Johnson, and James A. Robinson. "Reversal of Fortune: Geography and Institutions in the Making of the Modern World Income Distribution." Working Paper 8460, National Bureau of Economic Research, September 2001.

Adelson, Howard. "Early Medieval Trade Routes." *American Historical Review* 65, no. 2 (1960): 271–87.

Albrecht-Carrié, René. *A Diplomatic History of Europe since the Congress of Vienna*. New York: Harpers & Brothers, 1958.

Allan, J. A. "Virtual Water: A Strategic Resource." *Ground Water* 36, no. 4 (1998): 545–46.

Andrews-Speed, P., X. Liao, and R. Dannreuther. *The Strategic Implications of China's Energy Needs*. Adelphi Papers, no. 346. London: International Institute for Strategic Studies, 2002.

Angell, Norman. *The Great Illusion*. New York: Putnam's Sons, 1934.

Anisimov, E. V. "The Imperial Heritage of Peter the Great in the Foreign Policy of His Early Successors." In *Imperial Russian Foreign Policy*, edited by Hugh Ragsdale, 22–24. Cambridge: Cambridge University Press, 1993.

Aron, Raymond. *Peace and War*. Garden City, NY: Doubleday, 1966.

Atwell, William. "Ming China and the Emerging World Economy, c. 1470–1650." In *The Cambridge History of China*, vol. 8, *The Ming Dynasty, 1368–1644*, edited by Frederick W. Mote and Denis Twitchett, pt. 2:376–416. New York: Cambridge University Press, 1998.

———. "A Seventeenth-Century 'General Crisis' in East Asia?" *Modern Asian Studies* 24, no. 4 (1990): 661–82.

———. "Some Observations on the 'Seventeenth-Century Crisis' in China and Japan." *Journal of Asian Studies* 45, no. 2 (1986): 223–44.

Babinger, Franz. *Mehmed the Conqueror and His Time*. Princeton, NJ: Princeton University Press, 1978.

Baechler, J., J. Hall, and M. Mann, eds. *Europe and the Rise of Capitalism.* New York: Basil Blackwell, 1988.

Banlaoi, Rommel. "Maritime Terrorism in Southeast Asia." *Naval War College Review* 58, no. 4 (2005): 63–80.

Barbaro, Nicolò. *Diary of the Siege of Constantinople.* New York: Exposition Press, 1969.

Barker, Thomas. *Double Eagle and Crescent.* Albany: State University of New York Press, 1967.

Barkey, Karen. *Bandits and Bureaucrats.* Ithaca, NY: Cornell University Press, 1997.

Baron, Hans. "The Anti-Florentine Discourse of the Doge Tommaso Mocenigo (1414–23): Their Date and Partial Forgery." *Speculum* 27, no. 3 (1952): 323–42.

———. "A Struggle for Liberty in the Renaissance: Florence, Venice, and Milan in the Early Quattrocento—Part One." *American Historical Review* 58, no. 2 (1953): 265–89.

———. "A Struggle for Liberty in the Renaissance: Florence, Venice, and Milan in the Early Quattrocento—Part Two." *American Historical Review* 58, no. 3 (1953): 544–70.

Barron, Michael J. "China's Strategic Modernization: The Russian Connection." *Parameters* 31, no. 4 (2001–2): 72–86.

Beeching, Jack. *The Galleys at Lepanto.* London: Hutchinson, 1982.

Benzoni, Gino. "Venezia, ossia il mito modulato." *Studi Veneziani* 19 (1990): 15–33.

Bentley, Jerry. "Hemispheric Integration, 500–1500 C.E." *Journal of World History* 9, no. 2 (1998): 237–54.

Bergère, Marie-Claire. "On the Historical Origins of Chinese Underdevelopment." *Theory and Society* 13, no. 3 (1984): 327–38.

Bernard, Henri, S. J. "Les debuts des relations diplomatiques entre le Japon et les Espagnols des Iles Philippines (1571–1594)." *Monumenta Nipponica* 1, no. 1 (1938): 99–137.

Black, Jeremy. *Maps and History.* New Haven, CT: Yale University Press, 1997.

———. *War and the World: Military Power and the Fate of Continents, 1450–2000.* New Haven, CT: Yale University Press, 2000.

Blechman, Barry, and Robert Weinland. "Why Coaling Stations Are Necessary in the Nuclear Age." *International Security* 2, no. 1 (1977): 88–99.

Bloch, Marc. *Feudal Society.* London: Routledge & Kegan Paul, 1961.

Boulding, Kenneth. *Conflict and Defense.* New York: Harper & Row, 1963.

Bowman, Isaiah. *Geography in Relation to the Social Sciences.* New York: Charles Scribner's Sons, 1934.

———. "Geography vs. Geopolitics." *Geographical Review* 32, no. 4 (1942): 646–58.

Boxer, C. R. *Portuguese Conquest and Commerce in Southern Asia, 1500–1750.* London: Variorum Reprints, 1985.

———. *The Portuguese Seaborne Empire, 1415–1825.* London: Hutchinson, 1969.

Braudel, Fernand. *The Mediterranean and the Mediterranean World in the Age of Philip II.* Translated by Siân Reynolds. Vol. 1. 1972. Reprint, Berkeley and Los Angeles: University of California Press, 1995.

———. *The Mediterranean and the Mediterranean World in the Age of Philip II.* Translated by Siân Reynolds. Vol. 2. New York: Harper & Row, 1973.

Bravetta, Ettore. *Enrico Dandolo.* Milan: Casa Editrice "Alpes," 1929.

Bréhier, Louis. "Attempts at Reunion of the Greek and Latin Churches." In *The Cambridge Medieval History*, vol. 4, edited by J. R. Tanner, C. W. Previté-Orton, and Z. N. Brooke, 594–626. New York: Macmillan, 1926.

———. "The Greek Church: Its Relations with the West up to 1054." In *The Cambridge Medieval History*, vol. 4, edited by J. R. Tanner, C. W. Previté-Orton, and Z. N. Brooke, 246–73. New York: Macmillan, 1926.

Brodie, Bernard. "Strategy as a Science." *World Politics* 1, no. 4 (1949): 467–88.

Brooks, Stephen. "The Globalization of Production and the Changing Benefits of Conquest." *Journal of Conflict Resolution* 43, no. 5 (1999): 646–70.

Brown, Horatio. "Venice." In *The Cambridge Medieval History*, vol. 4, edited by J. R. Tanner, C. W. Previté-Orton, and Z. N. Brooke, 385–414. New York: Macmillan, 1926.

Bryer, Anthony. "Greek Historians on the Turks: The Case of the First Byzantine-Ottoman Marriage." In *The Writing of History in the Middle Ages*, edited by R. H. C. Davis and J. M. Wallace-Hadrill, 471–93. Oxford: Clarendon, 1981.

———. "Nicaea, a Byzantine City." *History Today* 21, no. 1 (1971): 22–31.

Brzezinski, Zbigniew. *Game Plan*. New York: Atlantic Monthly Press, 1986.

———. *The Grand Chessboard*. New York: HarperCollins, 1998.

Buck, David. "Was It Pluck or Luck That Made the West Grow Rich?" *Journal of World History* 10, no. 2 (1999): 413–30.

Burckhardt, Jacob. *The Civilization of the Renaissance in Italy*. London: Phaidon, 1955.

Burles, Mark, and Abram N. Shulsky. *Patterns in China's Use of Force: Evidence from History and Doctrinal Writings*. Santa Monica, CA: Rand, 2000.

Caetani, Onorato, and Gerolamo Diedo. *La battaglia di Lepanto*. Palermo: Sellerio, 1995.

Canosa, Romano. *Lepanto: Storia della "Lega Santa" contro i turchi*. Rome: Sapere, 2000.

Cecil, Lamar. "Coal for the Fleet That Had to Die." *American Historical Review* 69, no. 4 (1964): 990–1005.

Chalandon, Ferdinand. "The Earlier Comneni." In *The Cambridge Medieval History*, vol. 4, edited by J. R. Tanner, C. W. Previté-Orton, and Z. N. Brooke, 318–50. New York: Macmillan, 1926.

Chase, Robert, Emily Hill, and Paul Kennedy. *The Pivotal States*. New York: Norton, 1998.

Chaudhuri, K. N. *Asia before Europe: Economy and Civilisation of the Indian Ocean from the Rise of Islam to 1750*. Cambridge: Cambridge University Press, 1990.

Christensen, Thomas. "Posing Problems without Catching Up." *International Security* 45, no. 4 (2001): 5–40.

Christian, David. "The Case for 'Big History.'" *Journal of World History* 2, no. 2 (1991): 223–38.

———. "Silk Roads or Steppe Roads? The Silk Roads in World History." *Journal of World History* 11, no. 1 (2000): 1–26.

Cipolla, Carlo. *Guns, Sails, and Empires*. New York: Pantheon, 1965.

Citarella, Armand. "Patterns in Medieval Trade: The Commerce of Amalfi before the Crusades." *Journal of Economic History* 28, no. 4 (1968): 531–55.

Cizakca, Murat. "Price History and the Bursa Silk Industry: A Study in Ottoman Industrial Decline, 1550–1650." *Journal of Economic History* 40, no. 3 (1980): 533–50.

Cliff, Roger, and Jeremy Shapiro. "The Shift to Asia: Implications for U.S. Land Power." In *The U.S. Army and the New National Security Strategy,* edited by L. Davis and J. Shapiro. MR-1657-A. Santa Monica, CA: Rand, 2003.

Clokie, H. McD. "Geopolitics—New Super-Science or Old Art?" *Canadian Journal of Economics and Political Science* 10, no. 4 (1944): 492–502.

Coccon, Gino. *La Venezia di Terra.* Venice: Edizioni Helvetia, 1985.

Cohen, Saul. *Geography and Politics in a World Divided.* New York: Random House, 1963.

Cooper, Michael. "The Mechanics of the Macao-Nagasaki Silk Trade." *Monumenta Nipponica* 27, no. 4 (1972): 423–33.

Costantini, Massimo. *L'acqua di Venezia.* Venice: Arsenale Editrice, 1984.

Cot, André. *Suleiman the Magnificent.* London: Saqi Books, 1989.

Craig, Gordon A. "The Historian and the Study of International Relations." *American Historical Review* 88, no. 1 (1983): 1–11.

Cremonesi, Arduino. *La sfida turca contro gli Asburgo e Venezia.* Udine: Arti Grafiche Friulane, 1976.

Crowl, Philip. "Alfred Thayer Mahan: The Naval Historian." In *Makers of Modern Strategy,* edited by Peter Paret, 452–54. Princeton, NJ: Princeton University Press, 1986.

Currey, E. Hamilton. *Sea Wolves of the Mediterranean.* New York: Stokes, 1910.

Dardess, John W. "The Transformations of Messianic Revolt and the Founding of the Ming Dynasty." *Journal of Asian Studies* 29, no. 3 (1970): 539–58.

Daveluy, René. *La lutte pour l'empire de la mer.* Paris: Augustin Challamel, 1916.

David, Steven R. "Why the Third World Matters." *International Security* 14, no. 1 (1989): 50–85.

Dean, Trevor. "Venetian Economic Hegemony: The Case of Ferrara, 1220–1500." *Studi Veneziani* 12 (1986): 45–98.

De Booy, H. Th. "The Life Lines of the British Empire." *Pacific Affairs* 10, no. 2 (1937): 161–67.

Deng, Kent G. "A Critical Survey of Recent Research in Chinese Economic History." *Economic History Review* 53, no. 1 (2000): 1–28.

Desch, Michael. "The Keys That Lock Up the World." *International Security* 14, no. 1 (1989): 86–121.

———. *When the Third World Matters.* Baltimore: Johns Hopkins University Press, 1993.

Deudney, Daniel. *Whole Earth Security: A Geopolitics of Peace.* Worldwatch Paper no. 55. Washington, DC: Worldwatch Institute, July 1983.

Devries, Kelly. *Medieval Military Technology.* Peterborough, ON: Broadview, 1992.

Diamond, Jared. *Guns, Germs, and Steel.* New York: Norton, 1997.

Diehl, Charles. "The Fourth Crusade and the Latin Empire." In *The Cambridge Medieval History,* vol. 4, edited by J. R. Tanner, C. W. Previté-Orton, and Z. N. Brooke, 415–31. New York: Macmillan, 1926.

———. *La République de Venise.* Paris: Flammarion, 1985.

Diehl, Paul. "Contiguity and Military Escalation in Major Power Rivalries, 1816–1980." *Journal of Politics* 47, no. 4 (1985): 1203–11.

Dorpalean, Andreas. *The World of General Haushofer.* New York: Farrar & Rinehart, 1942.

Downing, Brian. *The Military Revolution and Political Change.* Princeton, NJ: Princeton University Press, 1992.

Doyle, Michael. *Empires.* Ithaca, NY: Cornell University Press, 1986.

Dreyer, Edward L. *Early Ming China: A Political History, 1355–1435.* Stanford, CA: Stanford University Press, 1982.

Dubbs, Homer, and Robert Smith. "Chinese in Mexico City in 1635." *Far Eastern Quarterly* 1, no. 4 (1942): 387–89.

Dugan, Arthur Butler. "Mackinder and His Critics Reconsidered." *Journal of Politics* 24, no. 2 (1962): 241–57.

Earle, Edward Mead. "Adam Smith, Alexander Hamilton, Friedrich List: The Economic Foundations of Military Power." In *Makers of Modern Strategy,* edited by Edward Mead Earle, 148–52. Princeton, NJ: Princeton University Press, 1943.

———. "Power Politics and American World Policy." *Political Science Quarterly* 58, no. 1 (1943): 94–106.

East, W. Gordon. *The Geography behind History.* 1965. New ed. New York: Norton, 1999.

———. *An Historical Geography of Europe.* New York: Dutton, 1935.

Elvin, Mark. *The Pattern of the Chinese Past.* Stanford, CA: Stanford University Press, 1973.

———. "Why China Failed to Create an Endogenous Industrial Capitalism: A Critique of Max Weber's Explanation." *Theory and Society* 13, no. 3 (1984): 379–91.

Fairbank, John K. "Tributary Trade and China's Relations with the West." *Far Eastern Quarterly* 1, no. 2 (1942): 129–49.

Fairbank, John K., Edwin O. Reischauer, and Albert Craig. *East Asia: Tradition and Transformation.* Boston: Houghton Mifflin, 1978.

Fairgrieve, James. *Geography and World Power.* London: University of London Press, 1932.

Farmer, Edward L. *Early Ming Government: The Evolution of Dual Capitals.* Cambridge, MA: Harvard University Press, 1976.

Faroqhi, Suraiya. "The Venetian Presence in the Ottoman Empire, 1600–30." In *The Ottoman Empire and the World Economy,* edited by Huri Islamoglu-Inan, 311–44. New York: Cambridge University Press, 1987.

Ferguson, Niall. *Colossus.* New York: Penguin, 2004.

Fettweis, Christopher J. "Sir Halford Mackinder, Geopolitics, and Policymaking in the 21st Century." *Parameters* 30, no. 2 (2000): 58–71.

Filesi, Teobaldo. *China and Africa in the Middle Ages.* London: Frank Cass, 1972.

Finlay, Robert. "Crisis and Crusade in the Mediterranean: Venice, Portugal, and the Cape Route to India (1498–1509)." *Studi Veneziani* 28 (1994): 45–90.

———. "Fabius Maximus in Venice: Doge Andrea Gritti, the War of Cambrai, and the Rise of Habsburg Hegemony, 1509–1530." *Renaissance Quarterly* 53, no. 4 (2000): 988–1031.

Fleet, Kate. *European and Islamic Trade in the Early Ottoman State.* Cambridge: Cambridge University Press, 1999.

Fletcher, Joseph. "China and Central Asia." In *The Chinese World Order,* edited by John K. Fairbank, 208–24. Cambridge, MA: Harvard University Press, 1968.

Flynn, Dennis O. "Comparing the Tokagawa Shogunate with Hapsburg Spain: Two Silver-

Based Empires in a Global Setting." In *The Political Economy of Merchant Empires,* edited by James D. Tracy, 332–59. Cambridge: Cambridge University Press, 1991.

Flynn, Dennis O., and Arturo Giraldez. "Spanish Profitability in the Pacific." In *Pacific Centuries,* edited by Dennis O. Flynn, Lionel Frost, and A. J. H. Latham, 23–37. New York: Routledge, 1999.

Frank, Andre Gunder. *ReOrient: Global Economy in the Asian Age.* Berkeley and Los Angeles: University of California Press, 1998.

Friedberg, Aaron. "11 September and the Future of Sino-American Relations." *Survival* 44, no. 1 (2002): 33–50.

——. *In the Shadow of the Garrison State.* Princeton, NJ: Princeton University Press, 2000.

——. "Is the United States Capable of Acting Strategically?" *Washington Quarterly,* winter 1991, 5–23.

——. "Ripe for Rivalry: Prospects for Peace in a Multipolar Asia." *International Security* 18, no. 3 (1993–94): 3–31.

——. *The Weary Titan.* Princeton, NJ: Princeton University Press, 1988.

——. "Will Europe's Past Be Asia's Future?" *Survival* 42, no. 3 (2000): 147–59.

Fry, Robert. "End of the Continental Century." *RUSI Journal* 143, no. 3 (1998): 15–18.

Fubini, Riccardo. "The Italian League and the Policy of the Balance of Power at the Accession of Lorenzo de' Medici." In "The Origins of the State in Italy, 1300–1600," edited by Julius Kirshner. *Journal of Modern History* 67, no. 4 (1995): S166–S199.

Fukuyama, Francis. "Re-Envisioning Asia." *Foreign Affairs* 84, no. 1 (2005): 75–87.

——. *State-Building: Governance and World Order in the 21st Century.* Ithaca, NY: Cornell University Press, 2004.

Fuller, William. *Strategy and Power in Russia (1600–1914).* New York: Free Press, 1992.

Furniss, Edgar. "The Contribution of Nicholas John Spykman to the Study of International Politics." *World Politics* 4, no. 3 (1952): 382–401.

Gaddis, John Lewis. *Strategies of Containment.* New York: Oxford University Press, 1982.

——. *The United States and the End of the Cold War.* New York: Oxford University Press, 1992.

Gallup, John Luke, and Jeffrey D. Sachs. "Location, Location: Geography and Economic Development." *Harvard International Review* 21, no. 1 (1998–99): 56–62.

Gallup, John Luke, Jeffrey D. Sachs, and Andrew D. Mellinger. "Geography and Economic Development." Working Paper W6849, National Bureau of Economic Research, December 1998.

Gerace, Michael P. "Between Mackinder and Spykman: Geopolitics, Containment, and After." *Comparative Strategy* 10 (1991): 347–64.

Gibbon, Edward. *The History of the Decline and Fall of the Roman Empire.* Edited by David Womersley. 3 vols. London: Penguin, 1994.

Gibbons, Herbert Adams. *The Foundation of the Ottoman Empire.* New York: Century, 1916.

Gilbert, Felix. "Machiavelli: The Renaissance of the Art of War." In *Makers of Modern Strategy,* edited by Peter Paret, 11–31. Princeton, NJ: Princeton University Press, 1986.

——. *The Pope, His Banker, and Venice.* Cambridge, MA: Harvard University Press, 1980.

Gilpin, Robert. *War and Change in World Politics.* Cambridge: Cambridge University Press, 1981.

Glauser, Charles, and Chaim Kaufmann. "What Is the Offense-Defense Balance and Can We Measure It?" *International Security* 22, no. 4 (1998): 44–82.

Godfrey, John. *1204: The Unholy Crusade.* Oxford: Oxford University Press, 1980.

Goldstein, Lyle, and William Murray. "Undersea Dragons: China's Maturing Submarine Force." *International Security* 28, no. 4 (2004): 161–96.

Goldstone, Jack. "East and West in the Seventeenth Century: Political Crises in Stuart England, Ottoman Turkey, and Ming China." *Comparative Studies in Society and History* 30, no. 1 (1988): 103–42.

Gooch, John. "The Weary Titan: Strategy and Policy in Great Britain, 1890–1918." In *The Making of Strategy*, edited by Williamson Murray, Macgregor Knox, and Alvin Bernstein, 278–306. New York: Cambridge University Press, 1994.

Goodwin, Godfrey. *The Janissaries.* London: Saqi Books, 1994.

Gottmann, Jean. "The Background of Geopolitics." *Military Affairs* 6, no. 4 (1942): 197–206.

———. "Geography and International Relations." *World Politics* 3, no. 2 (1951): 153–73.

Gray, Colin. "The Continued Primacy of Geography." *Orbis*, spring 1996, 247–59.

———. "Geography and Grand Strategy." *Comparative Strategy* 10 (1991) 311–29.

———. "Strategy in the Nuclear Age: The United States, 1945–1991." In *The Making of Strategy*, edited by Williamson Murray, Macgregor Knox, and Alvin Bernstein, 579–613. New York: Cambridge University Press, 1994.

Grubb, James. "When Myths Lose Power: Four Decades of Venetian Historiography." *Journal of Modern History* 58, no. 1 (1986): 43–94.

Guicciardini, Francesco. *Storie fiorentine.* Milan: TEA, 1991.

Guilmartin, John Francis. *Gunpowder and Galleys.* New York: Cambridge University Press, 1974.

Gungwu, Wang. "Ming Foreign Relations: Southeast Asia." In *The Cambridge History of China*, vol. 8, *The Ming Dynasty, 1368–1644*, edited by Frederick W. Mote and Denis Twitchett, pt. 2:301–32. New York: Cambridge University Press, 1998.

Gyorgy, Andrew. "The Application of German Geopolitics: Geo-Sciences." *American Political Science Review* 37, no. 4 (August 1943): 677–86.

———. "The Geopolitics of War: Total War and Geostrategy." *Journal of Politics* 5, no. 4 (1943): 347–62.

Hall, A. R. "Epilogue: The Rise of the West." In *A History of Technology*, vol. 3, edited by Charles Singer, E. J. Holmyard, A. R. Hall, and Trevor Williams, 709–21. Oxford: Clarendon, 1957

Hall, Kenneth R. "Economic History of Early Southeast Asia." In *The Cambridge History of Southeast Asia*, ed. Nicholas Tarling, 1:183–275. New York: Cambridge University Press, 1992.

Hamdani, Abbas. "Ottoman Response to the Discovery of America and the New Route to India." *Journal of the American Oriental Society* 101, no. 3 (1981): 323–30.

Hamilton, Alexander, James Madison, and John Jay. *The Federalist Papers.* Edited and with an introduction by Garry Wills. New York: Bantam Books, 1982.

Hamilton, Earl J. "The Role of Monopoly in the Overseas Expansion and Colonial Trade of Europe before 1800. *American Economic Review* 38, no. 2 (1948): 33–53.

Hansen, David. "The Immutable Importance of Geography." *Parameters* 27, no. 1 (1997): 55–64.

Haskins, Charles. *The Renaissance of the Twelfth Century.* Cambridge, MA: Harvard University Press, 1927.

Hausmann, Ricardo. "Prisoners of Geography." *Foreign Policy,* no. 122 (January–February 2001): 44–53.

Henrikson, Alan. "The Geographical 'Mental Maps' of American Foreign Policy Makers." *International Political Science Review* 1, no. 4 (1980): 495–530.

———. "The Map as an 'Idea': The Role of Cartographic Imagery during the Second World War." *American Cartographer* 2, no. 1 (1975): 19–53.

Herwig, Holger H. "*Geopolitik:* Haushofer, Hitler, and Lebensraum." *Journal of Strategic Studies* 22, nos. 2–3 (1999): 218–41.

Herz, John. "Idealist Internationalism and the Security Dilemma." *World Politics* 2, no. 2 (1950): 157–80.

Hess, Andrew. "The Evolution of the Ottoman Seaborne Empire in the Age of the Oceanic Discoveries, 1453–1525." *American Historical Review* 75, no. 7 (1970): 1892–1919.

Hirschman, Albert. *The Strategy of Economic Development.* New Haven, CT: Yale University Press, 1958.

Hocquet, Jean-Claude. *Le sel et la fortune de Venise.* 2 vols. Villeneuve D'Ascq: Presses Universitaires de Lille, 1978–79.

Hodgson, Francis C. *The Early History of Venice.* London: George Allen, 1901.

Hoffman, Stanley. "An American Social Science: International Relations." *Daedalus* 106, no. 3 (1977): 41–60.

Hoffmann, J. Wesley. "The Commerce of the German Alpine Passes during the Early Middle Ages." *Journal of Political Economy* 31, no. 6 (1923): 826–36.

Hoover, Edgar. *The Location of Economic Activity.* New York: McGraw-Hill, 1948.

Hopf, Ted. "Polarity, the Offense-Defense Balance, and War." *American Political Science Review* 85, no. 2 (1991): 475–93.

Horniker, Arthur Leon. "The Corps of the Janizaries." *Military Affairs* 8, no. 3 (1944): 177–204.

Hough, Richard. *The Fleet That Had to Die.* New York: Viking, 1958.

Howard, Michael. "The Forgotten Dimensions of Strategy." *Foreign Affairs* 57, no. 5 (1979): 975–86.

———. "The Influence of Geopolitics on the East-West Struggle." *Parameters* 18, no. 3 (1988): 13–17.

———. "The Military Factor in European Expansion." In *The Expansion of International Society,* edited by Hedley Bull and Adam Watson, 33–42. New York: Oxford University Press, 1986.

———. "The Relevance of Traditional Strategy." *Foreign Affairs* 51, no. 2 (1973): 253–66.

"How to Shorten Tanker Voyages." *Oil and Gas Journal* 90, no. 52 (1992): 34.

Hucker, Charles. "Hu Tsung-hsien's Campaign against Hsu Hai, 1556." In *Chinese Ways in Warfare,* edited by Frank Kierman and John K. Fairbank, 273–307. Cambridge, MA: Harvard University Press, 1974.

——. *The Ming Dynasty, Its Origins and Evolving Institutions.* Ann Arbor: Center for Chinese Studies, University of Michigan, 1978.

——. *The Traditional Chinese State in Ming Times (1368–1644).* Tucson: University of Arizona Press, 1961.

Hudson, G. F. *Europe and China.* London: Edward Arnold, 1931.

Hurewitz, J. C. "Russia and the Turkish Straits: A Revaluation of the Origins of the Problem." *World Politics* 14, no. 4 (1962): 605–32.

Inalcik, Halil. *Essays in Ottoman History.* Istanbul: Eren, 1998.

——. "Mehmed the Conqueror (1432–1481) and His Time." *Speculum* 35, no. 3 (1960): 408–27.

——. *The Ottoman Empire: The Classical Age, 1300–1600.* London: Weidenfeld & Nicolson, 1973.

Inalcik, Halil, and Cemal Kafadar, eds. *Suleyman the Second and His Time.* Istanbul: ISIS, 1993.

Jelavich, Barbara. *The Ottoman Empire, the Great Powers, and the Straits Question, 1870–1887.* Bloomington: Indiana University Press, 1973.

Jervis, Robert. "Cooperation under the Security Dilemma." *World Politics* 30, no. 2 (1978): 167–214.

——. "From Balance to Concert: A Study of International Security Cooperation." *World Politics* 38, no. 1 (1985): 58–79.

——. "Hypotheses on Misperception." *World Politics* 20, no. 3 (1968): 454–79.

Johnston, Alastair Iain. *Cultural Realism: Strategic Culture and Grand Strategy in Chinese History.* Princeton, NJ: Princeton University Press, 1995.

Jones, Eric. "Disasters and Economic Differentiation across Eurasia: A Reply." *Journal of Economic History* 45, no. 3 (1985): 675–82.

——. "A Framework for the History of Economic Growth of Southeast Asia." *Australian Economic History Review* 31, no. 1 (1991): 3–19.

——. "The Real Question about China: Why Was the Song Economic Achievement Not Repeated?" *Australian Economic History Review* 30, no. 2 (1990): 5–22.

Jones, Eric, Lionel Frost, and Colin White. *Coming Full Circle.* Boulder, CO: Westview, 1993.

Jones, Stephen. "The Power Inventory and National Strategy." *World Politics* 6, no. 4 (1954): 421–52.

——. "Global Strategic Views." *Geographical Review* 45, no. 4 (1955): 492–508.

Josephus, Flavius. *The Jewish War.* Translated by G. A. Williamson. Rev. ed., with new introduction, notes, and appendixes by E. Mary Smallwood. New York: Penguin, 1981.

Kafadar, Cemal. *Between Two Worlds: The Construction of the Ottoman State.* Berkeley and Los Angeles: University of California Press, 1995.

Kang, David C. "Hierarchy in Asian International Relations: 1300–1900." *Asian Security* 1, no. 1 (2005): 53–79.

Karpov, Sergei Pavlovic. *L'impero di Trebisonda.* Rome: Il Veltro Editrice, 1986.

Katele, Irene. "Piracy and the Venetian State: The Dilemma of Maritime Defense in the Fourteenth Century." *Speculum* 63, no. 4 (1988): 865–89.

Kaufman, Robert. "'To Balance or to Bandwagon?' Alignment Decisions in 1930s Europe." *Security Studies* 1, no. 3 (1992): 417–47.

Kaysen, Carl. "Is War Obsolete? A Review Essay." *International Security* 14, no. 4 (1990): 42–64.

Kemp, Geoffrey, and Robert Harkav. *Strategic Geography of the Changing Middle East.* Washington, DC: Carnegie Endowment for International Peace, 1992.

Kennedy, Dane. "The Expansion of Europe." *Journal of Modern History* 59, no. 2 1987): 331–43.

Kennedy, Paul. *The Rise and Fall of British Naval Mastery.* Amherst, NY: Humanity Books, 1998.

———. *The Rise and Fall of the Great Powers.* New York: Random House, 1987.

———. *The Rise of the Anglo-German Antagonism, 1860–1914.* London: Ashfield, 1980.

Keohane, Robert, ed. *Neorealism and Its Critics.* New York: Columbia University Press, 1986.

Kinross, Lord. *The Ottoman Centuries.* New York: William Morrow, 1977.

Kiss, George. "Political Geography into Geopolitics." *Geographical Review* 32, no. 4 (1942): 632–45.

Kissinger, Henry. "Strategy and Organization." *Foreign Affairs* 35, no. 3 (1957): 379–94.

Knorr, Klaus. *The Power of Nations.* New York: Basic Books, 1975.

"Knowing Your Place." *Economist,* 13 March 1999, 92.

Konigsberg, Charles. "Climate and Society: A Review of the Literature." *Journal of Conflict Resolution* 4, no. 1 (1960): 67–82.

Köprülü, M. Fuad. *The Origins of the Ottoman Empire.* Albany: State University of New York Press, 1992.

Krepinevich, Andrew. Testimony before the Senate Armed Services AirLand Subcommittee (10 March 1999). www.csbaonline.org/4Publications/Archive/T.19990310.The_Future_Of_Tact/T.19990310.The_Future_Of_Tact.htm

Kristof, Ladis. "The Nature of Frontiers and Boundaries." *Annals of the Association of American Geographers* 49, no. 3 (1959): 269–82.

———. "The Origins and Evolution of Geopolitics." *Journal of Conflict Resolution* 4, no. 1 (1960): 15–51.

Krugman, Paul. *Geography and Trade.* Cambridge, MA: MIT Press, 1991.

———. "Increasing Returns and Economic Geography." *Journal of Political Economy* 99, no. 3 (1991): 483–99.

Kruszewski, Charles. "International Affairs: Germany's Lebensraum." *American Political Science Review* 34, no. 5 (1940): 964–75.

Lam, Truong Buu. "Intervention versus Tribute in Sino-Vietnamese Relations, 1788–1790." In *The Chinese World Order,* edited by John K. Fairbank, 165–75. Cambridge, MA: Harvard University Press, 1968.

Lamb, Alastair. *Asian Frontiers.* New York: Praeger, 1968.

Lambeth, Benjamin. "Air Power, Space Power, and Geography." *Journal of Strategic Studies* 22, nos. 2–3 (1999): 63–82.

Landes, David S. *The Wealth and Poverty of Nations.* New York: Norton, 1998.

———. "What Room for Accident in History? Explaining Big Changes by Small Events." *Economic History Review* 47, no. 4 (1994): 637–56.

Lane, Frederic C. "The Economic Meaning of the Invention of the Compass." *American Historical Review* 68, no. 3 (1963): 605–17.

——. "Pepper Prices before Da Gama." *Journal of Economic History* 28, no. 4 (1968): 590–97.

——. "Recent Studies on the Economic History of Venice." *Journal of Economic History* 23, no. 3 (1963): 312–34.

——. "Venetian Shipping during the Commercial Revolution." *American Historical Review* 38, no. 2 (1933): 210–39.

——. *Venetian Ships and Shipbuilders of the Renaissance.* Baltimore: Johns Hopkins University Press, 1934.

——. *Venice: A Maritime Republic.* Baltimore: Johns Hopkins University Press, 1973.

Langer, William, and Robert Blake. "The Rise of the Ottoman Turks and Its Historical Background." *American Historical Review* 37, no. 3 (1932): 468–505.

Lattimore, Owen. "Inner Asian Frontiers: Chinese and Russian Margins of Expansion." *Journal of Economic History* 7, no. 1 (1947): 24–52.

——. *Inner Asian Frontiers of China.* New York: American Geographical Society, 1940.

Lesure, Michel. *Lépante: La crise de l'Empire ottoman.* Paris: Juillard, 1972.

Levathes, Louise. *When China Ruled the Seas.* New York: Simon & Schuster, 1994.

Levy, Jack. "The Offensive/Defensive Balance of Military Technology: A Theoretical and Historical Analysis." *International Studies Quarterly* 38, no. 2 (1984): 219–38.

Lewis, Archibald. "The Closing of the Mediaeval Frontier, 1250–1350." *Speculum* 33, no. 4 (1958): 475–83.

——. "The Islamic World and the Latin West, 1350–1500." *Speculum* 65, no. 4 (1990): 833–44.

Lewis, Bernard. *The Muslim Discovery of Europe.* 1982. Reprint, New York: Norton, 2001.

Libby, Lester J., Jr. "Venetian History and Political Thought after 1509." *Studies in the Renaissance* 20 (1973): 7–45.

Liberman, Peter. *Does Conquest Pay?* Princeton, NJ: Princeton University Press, 1996.

——. "The Spoils of Conquest." *International Security* 18, no. 2 (1993): 125–53.

Libicki, Martin. "The Emerging Primacy of Information." *Orbis,* spring 1996, 261–76.

Lieven, Dominic. *Empire.* London: John Murray, 2000.

Lindner, Rudi Paul. *Nomads and Ottomans in Medieval Anatolia.* Bloomington: Indiana University Research Institute for Inner Asian Studies, 1983.

Lindsay, Jack. *Byzantium into Europe.* London: Bodley Head, 1952.

Lo, Jung-Pang. "The Emergence of China as a Sea Power." *Far Eastern Quarterly* 14, no. 4 (1955): 489–503.

Lopez, Robert Sabatino. "European Merchants in the Medieval Indies: The Evidence of Commercial Documents." *Journal of Economic History* 3, no. 2 (1943): 164–84.

——. "Market Expansion: The Case of Genoa." *Journal of Economic History* 24, no. 4 (1964): 445–64.

Lorge, Peter. "War and Warfare in China, 1450–1815." In *War in the Early Modern World,* edited by Jeremy Black, 87–104. London: University College London Press, 1999.

Luzzati, Michele. "La dinamica secolare di un 'modello italiano.'" In *Storia dell'economia italiana,* edited by Ruggiero Romano, vol. 1. Turin: Einaudi, 1990.

Luzzatto, Gino. *Storia economica di Venezia dall'XI al XVI secolo.* Venice: Marsilio Editori, 1995.

Ma, Debin. "The Great Silk Exchange." In *Pacific Centuries,* edited by Dennis O. Flynn, Lionel Frost, and A. J. H. Latham, 38–65. London: Routledge, 1999.

Machiavelli, Niccolò. *Discourses.* New York: Modern Library, 1950.

———. *Il principe e altre opere politiche.* Italy: Garzanti, 1976.

Mackinder, Halford. *Democratic Ideals and Reality.* Washington, DC: National Defense University Press, 1996.

———. "The Geographical Pivot of History." *Geographical Journal* 23, no. 4 (1904): 421–39. Reprinted in *Democratic Ideals and Reality,* by Halford Mackinder, 175–93. Washington, DC: National Defense University Press, 1996.

———. "The Scope and Methods of Geography." *Proceedings of the Royal Geographical Society* 9 (1887): 141–60. Reprinted in *Democratic Ideals and Reality,* by Halford Mackinder, 151–73. Washington, DC: National Defense University Press, 1996.

Madden, Thomas. *A Concise History of the Crusades.* Oxford: Rowman & Littlefield, 1999.

———. "Venice's Hostage Crisis: Diplomatic Efforts to Secure Peace with Byzantium between 1171 and 1184." In *Medieval and Renaissance Venice,* edited by Ellen Kittell and Thomas Madden, 96–108. Urbana: University of Illinois Press, 1999.

Mahan, A. T. *The Influence of Sea Power upon History, 1660–1783.* 1890. Reprint, New York: Dover, 1987.

———. *The Interest of America in Sea Power, Present and Future.* 1897. Reprint, Boston: Little, Brown, 1918.

Mallett, M. E., and J. R. Hale. *The Military Organization of a Renaissance State.* Cambridge: Cambridge University Press, 1984.

Małowist, Marian. *Europa i jej ekspansja.* Warsaw: PWN, 1993.

Mancall, Mark. "The Ch'ing Tribute System: An Interpretive Essay." In *The Chinese World Order,* edited by John K. Fairbank, 63–89. Cambridge, MA: Harvard University Press, 1968.

Marshall, P. J. "Western Arms in Maritime Asia in the Early Phases of Expansion." *Modern Asian Studies* 14, no. 1 (1980): 13–28.

Mazzarella, Salvatore. Introduction to *La battaglia di Lepanto,* by Onorato Caetani and Gerolamo Diedo. Palermo: Sellerio, 1995.

McCarthy, Justin. *The Ottoman Turks.* London: Longman, 1997.

McGowan, Bruce. "The Middle Danube *cul-de-sac.*" In *The Ottoman Empire and the World Economy,* edited by Huri Islamoglu-Inan. Cambridge: Cambridge University Press, 1987.

McNeill, William. "The Changing Shape of World History." *History and Theory* 34, no. 2 (1995): 8–26.

———. "The Eccentricity of Wheels, or Eurasian Transportation in Historical Perspective." *American Historical Review* 92, no. 5 (1987): 1111–26.

———. *The Pursuit of Power.* Chicago: University of Chicago Press, 1982.

———. *The Rise of the West.* Chicago: University of Chicago Press, 1963.

———. *Venice: The Hinge of Europe.* Chicago: University of Chicago Press, 1974.

———. "World History and the Rise and Fall of the West." *Journal of World History* 9, no. 2 (1998): 215–36.

Mead, Walter Russell. *Power, Terror, Peace, and War.* New York: Vintage Books, 2005.

Moczulski, Leszek. *Geopolityka.* Warsaw: Dom Wydawniczy Bellona, 1999.

Montesquieu. *Considerations on the Causes of the Greatness of the Romans and Their Decline.* Indianapolis: Hackett, 1965.

Moodie, A. E. *Geography behind Politics.* London: Hutchison's University Library, 1947.

Morgenthau, Hans. *In Defense of the National Interest.* New York: Knopf, 1951.

———. *Politics among Nations.* New York: Knopf, 1948.

Most, Benjamin, and Harvey Starr. "Diffusion, Reinforcement, Geopolitics, and the Spread of War." *American Political Science Review* 74, no. 4 (1980): 932–46.

Mote, Frederick W. *Imperial China, 900–1800.* Cambridge, MA: Harvard University Press, 1999.

———. "The T'u-mu Incident of 1449." In *Chinese Ways in Warfare,* edited by Frank Kierman and John K. Fairbank, 243–72. Cambridge, MA: Harvard University Press, 1974.

Mueller, John. *Retreat from Doomsday: The Obsolescence of Major War.* New York: Basic Books, 1989.

Mungello, D. E. *The Great Encounter of China and the West, 1500–1800.* Lanham, MD: Rowman & Littlefield, 1999.

Munson, William. *The Last Crusade.* Dubuque, IA: Brown Book Company, 1969.

Murray, Williamson. "Some Thoughts on War and Geography." *Journal of Strategic Studies* 22, nos. 2–3 (1999): 201–17.

Nelson, Richard, and Gavin Wright. "The Rise and Fall of American Technological Leadership: The Postwar Era in Historical Perspective." *Journal of Economic Literature* 30, no. 4 (1992): 1931–64.

Neustadt, R., and E. May. *Thinking in Time.* New York: Free Press, 1986.

Niebuhr, Reinhold. *The Structure of Nations and Empires.* New York: Scribner, 1959.

Norwich, John Julius. *Venice: The Rise to Empire.* London: Allen Lane, 1977.

Nye, Joseph. "The Changing Nature of World Power." *Political Science Quarterly* 105, no. 2 (1990): 177–92.

Odell, Peter. *Oil and World Power.* New York: Taplinger, 1972.

Okey, Thomas. *The Story of Venice.* London: Dent & Sons, 1931.

O'Malley, William D. "Central Asia and South Caucasus as an Area of Operations: Challenges and Constraints." In *Faultlines of Conflict in Central Asia and the South Caucasus: Implications for the U.S. Army,* edited by O. Oliker and T. Szayna, chap. 8 MR-1598-A. Santa Monica, CA: Rand, 2003.

Orme, John. "The Utility of Force in a World of Scarcity." *International Security* 22, no. 3 (1997–98): 138–67.

Owens, Mackubin Thomas. "In Defense of Classical Geopolitics." *Naval War College Review* 52, no. 4 (1999): 59–76.

Pajewski, Janusz. *Buńczuk i koncerz.* Warsaw: Wiedza Powszechna, 1978.

Palmer, Alan. *The Decline and Fall of the Ottoman Empire.* London: John Murray, 1992.

Panikkar, K. M. *Asia and Western Dominance: A Survey of the Vasco da Gama Epoch of Asian History, 1498–1945*. London: Allen & Unwin, 1953.

Parker, Geoffrey. "Europe and the Wider World, 1500–1750: The Military Balance." In *The Political Economy of Merchant Empires*, edited by James D. Tracy, 161–95. Cambridge: Cambridge University Press, 1991.

———. *The Grand Strategy of Philip II*. New Haven, CT: Yale University Press, 1998.

———. "The Making of Strategy in Habsburg Spain: Philip II's 'Bid for Mastery,' 1556–1598." In *The Making of Strategy*, edited by Williamson Murray, Macgregor Knox, and Alvin Bernstein, 115–50. Cambridge: Cambridge University Press, 1994.

———. *The Military Revolution: Military Innovation and the Rise of the West, 1500–1800*. Cambridge: Cambridge University Press, 1996.

———. *Success Is Never Final*. New York: Basic Books, 2002.

Parry, J. H. *The Age of Reconnaissance*. New York: Mentor Books, 1964.

———. *The Discovery of the Sea*. New York: Dial, 1974.

———. *The Establishment of the European Hegemony, 1415–1715*. New York: Harper & Brothers, 1961.

———. *The Spanish Seaborne Empire*. New York: Knopf, 1977.

———. "Transport and Trade Routes." In *The Cambridge Economic History of Europe*, edited by E. E. Rich and C. H. Wilson, 4:155–222. Cambridge: Cambridge University Press, 1967.

Paul, T. V. " States, Security Function, and the New Global Forces." In *The Nation-State in Question*, edited by T. V. Paul, G. John Ikenberry, and John Hall, 139–65. Princeton, NJ: Princeton University Press, 2003.

Pears, Edwin. "The Ottoman Turks to the Fall of Constantinople." In *The Cambridge Medieval History*, vol. 4, edited by J. R. Tanner, C. W. Previté-Orton, and Z. N. Brooke, 653–705. New York: Macmillan, 1926.

Pearson, Frederic. "Geographic Proximity and Foreign Military Intervention." *Journal of Conflict Resolution* 18, no. 3 (1974): 432–60.

Pelissier, Roger. *The Awakening of China, 1793–1949*. London: Secker & Warburg, 1963.

Pertusi, Agostino. *Saggi veneto-bizantini*. Florence: Leo S. Olschki Editore, 1990.

Pirenne, Henri. *Medieval Cities: Their Origins and the Revival of Trade*. Princeton, NJ: Princeton University Press, 1952.

Pomeranz, Kenneth. *The Great Divergence: China, Europe, and the Making of the Modern World Economy*. Princeton, NJ: Princeton University Press, 2000.

Porciatti, Anna Maria. *Dall'impero ottomano alla nuova Turchia*. Florence: Alinea Editrice, 1997.

Posen, Barry R. "Command of the Commons." *International Security* 28, no. 1 (2003): 5–46.

Potkowski, Edward. *Warna 1444*. Warsaw: Wydawnictwo Bellona, 1990.

Prakash, Om. "The Portuguese and the Dutch in Asian Maritime Trade: A Comparative Analysis." In *Merchants, Companies, and Trade*, edited by Sushil Chaudhury and Michel Morineau, 175–88. Cambridge: Cambridge University Press, 1999.

Pryor, John. *Geography, Technology, and War*. Cambridge: Cambridge University Press, 1988.

———. "The Venetian Fleet for the Fourth Crusade and the Diversion of the Crusade to

Constantinople." In *The Experience of Crusading,* vol. 1, *Western Approaches,* edited by M. Bull and N. Housley, 103–23. Cambridge: Cambridge University Press, 2003.

Queller, Donald. *The Fourth Crusade.* Leicester: Leicester University Press, 1978.

Queller, Donald, and Gerald Day. "Some Arguments in Defense of the Venetians on the Fourth Crusade." *American Historical Review* 81, no. 4 (1976): 717–37.

Queller, Donald, and Irene Katele. "Venice and the Conquest of the Latin Kingdom of Jerusalem." *Studi Veneziani* 12 (1986): 15–43.

Ranke, Leopold von. *The Ottoman and the Spanish Empires.* London: Whittaker, 1975.

———. *Venezia nel Cinquecento.* Rome: Istituto della Enciclopedia Italiana, 1974.

Rapp, Richard T. "The Unmaking of the Mediterranean Trade Hegemony: International Trade Rivalry and the Commercial Revolution." *Journal of Economic History* 35, no. 3 (1975): 499–525.

Raudzens, George. "Military Revolution or Maritime Evolution? Military Superiorities or Transportation Advantages as Main Causes of European Colonial Conquests to 1788." *Journal of Military History* 63, no. 3 (1999): 631–41.

Ravegnani, Giorgio. "La Romania veneziana." In *Storia di Venezia dalle origini alla caduta della Serenissima: L'età del Comune,* edited by G. Cracco and G. Ortalli, 2:183–231. Rome: Istituto della Enciclopedia Italiana, 1995.

———. "Tra i due imperi." In *Storia di Venezia,* edited by Giorgio Cracco and Gherardo Ortalli, 2:33–80. Rome: Istituto della Enciclopedia Italiana, 1991.

Reid, Anthony. "The Seventeenth-Century Crisis in Southeast Asia." *Modern Asian Studies* 24, no. 4 (1990): 639–59.

Reischauer, Edwin O. "Notes on T'ang Dynasty Sea Routes." *Harvard Journal of Asiatic Studies* 5, no. 2 (1940): 142–64.

Rhodes, Edward. "' . . . From the Sea' and Back Again." *Naval War College Review* 52, no. 2 (1999): 13–54. www.nwc.navy.mil/press/Review/1999/spring/art1-sp9.htm.

Richards, John. "The Seventeenth-Century Crisis in South Asia." *Modern Asian Studies* 24, no. 4 (1990): 625–38.

Riley-Smith, Jonathan. *The Crusades.* New Haven, CT: Yale University Press, 1987.

Rose, Gideon. "Neoclassical Realism and Theories of Foreign Policy." *World Politics* 51, no. 1 (1998): 144–72.

Rosecrance, Richard. *The Rise of the Trading State.* New York: Basic Books, 1986.

———. *International Relations: Peace or War?* New York: McGraw-Hill, 1973.

Ross, Robert. "The Geography of Peace: East Asia in the Twenty-first Century." *International Security* 23, no. 4 (1999): 81–118.

Rossabi, Morris. "The 'Decline' of the Central Asian Caravan Trade." In *The Rise of Merchant Empires,* edited by James D. Tracy, 351–70. Cambridge: Cambridge University Press, 1990.

———. "The Ming and Inner Asia." In *The Cambridge History of China,* vol. 8, *The Ming Dynasty, 1368–1644,* edited by Frederick W. Mote and Denis Twitchett, pt. 2:221–71. New York: Cambridge University Press, 1998.

Roth, Gertraude. "The Manchu-Chinese Relationship, 1618–1636." In *From Ming to Ch'ing: Conquest, Region, and Continuity in Seventeenth-Century China,* edited by Jonathan Spence and John Wills, 2–38. New Haven, CT: Yale University Press, 1979.

Rothenberg, Gunther. "Aventinus and the Defense of the Empire against the Turks." *Studies in the Renaissance* 10 (1963): 60–67.

Runciman, Steven. *A History of the Crusades.* Vols. 2 and 3. Harmondsworth, UK: Penguin Books, 1954.

Sabattini, Mario, and Paolo Santangelo. *Storia della Cina.* Bari: Editori Laterza, 1986.

Sachs, Jeffrey D. "Nature, Nurture, and Growth." *Economist,* 14 June 1997, 19.

Sakai, Robert K. "The Ryuku (Liu-Ch'iu) Islands as a Fief of Satsuma." In *The Chinese World Order,* edited by John K. Fairbank, 112–34. Cambridge, MA: Harvard University Press, 1968.

Scammell, G. V. "After Da Gama: Europe and Asia since 1498." *Modern Asian Studies* 34, no. 3 (2000): 513–43.

———. "England, Portugal, and the *Estado da India,* c. 1500–1635." *Modern Asian Studies* 16, no. 2 (1982): 177–92.

———. *The First Imperial Age: European Overseas Expansion, 1400–1715.* New York: Routledge, 1990.

———. "Indigenous Assistance in the Establishment of Portuguese Power in Asia in the Sixteenth Century." *Modern Asian Studies* 14, no. 1 (1980): 1–11.

Schwoebel, Robert. "Coexistence, Conversion, and the Crusade against the Turks." *Studies in the Renaissance* 12 (1965): 164–87.

———. *The Shadow of the Crescent: The Renaissance Image of the Turk (1453–1517).* Nieuwkoop, Netherlands: B. de Graaf, 1967.

Sédillot, René. *Histoire du pétrole.* Paris: Fayard, 1974.

Serruys, Henry. "Mongols Ennobled during the Early Ming." *Harvard Journal of Asiatic Studies* 22 (December 1959): 209–60.

Seton-Watson, Hugh. *The Russian Empire: 1801–1917.* Oxford: Oxford University Press, 1967.

Shambaugh, David. "China Engages Asia: Reshaping the Regional Order." *International Security* 29, no. 3 (2004–5): 64–99.

Shaw, Stanford. *History of the Ottoman Empire and Modern Turkey.* Vol. 1. Cambridge: Cambridge University Press, 1976.

Simon, Sheldon W. "Is There a U.S. Strategy for East Asia?" *Contemporary Southeast Asia* 21, no. 3 (1999): 325–43.

Singer, J. David. "The Geography of Conflict: Introduction." *Journal of Conflict Resolution* 4, no. 1 (1960): 1–3.

Skinner, G. William. "Presidential Address: The Structure of Chinese History." *Journal of Asian Studies* 44, no. 2 (1985): 271–92.

Sloan, Geoffrey. "Sir Halford Mackinder: The Heartland Theory Then and Now." *Journal of Strategic Studies* 22, nos. 2–3 (1999): 15–38.

Smith, J. Russell. "The World Entrepôt." *Journal of Political Economy* 18, no. 9 (1910): 697–713.

Snyder, Glenn H. "Alliances, Balance, and Stability." *International Organization* 45, no. 1 (1991): 121–42.

So, Kwan-wai. *Japanese Piracy in Ming China during the 16th Century.* East Lansing: Michigan State University Press, 1975.

Sobieski, Jan. *Jan Sobieski: Listy do Marysieńki.* Edited by Leszek Kukulski. Warsaw: Czytelnik, 1970.

Speier, Hans. "Magic Geography." *Social Research* 8, no. 3 (1941): 310–30.

Spence, Jonathan. *The Chan's Great Continent.* New York: Norton, 1998.

———. *The Memory Palace of Matteo Ricci.* New York: Viking Penguin, 1984.

Sprout, Harold. "America's Strategy in World Politics." *American Political Science Review* 36, no. 5 (1942): 956–58.

———. "Frontiers of Defense." *Military Affairs* 5, no. 4 (1941): 217–21.

———. "Geopolitical Hypotheses in Technological Perspective." *World Politics* 15, no. 2 (1963): 187–212.

Sprout, Harold, and Margaret Sprout. *Foundations of International Politics.* Princeton, NJ: Van Nostrand, 1962.

———. "Geography and International Politics in an Era of Revolutionary Change." *Journal of Conflict Resolution* 4, no. 1 (1960): 145–61.

———. *Man-Milieu Relationship: Hypotheses in the Context of International Politics.* Princeton, NJ: Center of International Studies, 1956.

———. *Toward a Politics of the Planet Earth.* New York: Van Nostrand Reinhold, 1971.

Sprout, Margaret. "Mahan: Evangelist of Sea Power." In *Makers of Modern Strategy,* edited by Edward Mead Earle, 415–45. Princeton, NJ: Princeton University Press, 1943.

Spykman, Nicholas. *America's Strategy in World Politics.* 1942. Reprint, Hamden, CT: Archon Books, 1970.

———. "Frontiers, Security, and International Organization." *Geographical Review* 32, no. 3 (1942): 436–47.

———. "Geography and Foreign Policy, I." *American Political Science Review* 32, no. 1 (1938): 28–50.

———. "Geography and Foreign Policy, II." *American Political Science Review* 32, no. 2 (1938): 213–36.

———. *The Geography of the Peace.* New York: Harcourt, Brace, 1944.

Spykman, Nicholas, and Abbie Rollins. "Geographic Objectives in Foreign Policy, I." *American Political Science Review* 33, no. 3 (1939): 391–410.

———. "Geographic Objectives in Foreign Policy, II." *American Political Science Review* 33, no. 4 (1939): 591–614.

Starr, Harvey, and Benjamin Most. "The Substance and Study of Borders in International Relations Research." *International Studies Quarterly* 20, no. 4 (1976): 581–620.

St. Augustine. *The City of God.* Translated by Marcus Dools. Introduction by Thomas Merton. New York: Modern Library, 2000.

Steensgaard, Niels. "The Seventeenth-Century Crisis and the Unity of Eurasian History." *Modern Asian Studies* 24, no. 4 (1990): 683–97.

Stokes, Gale. "The Fates of Human Societies: A Review of Recent Macro-histories." *American Historical Review* 106, no. 2 (2001): 509–25.

———. "Why the West?" *Lingua Franca* 11, no. 8 (2001): 30–38.

Stokes, Mark A. *China's Strategic Modernization: Implications for the United States.* Carlisle Barracks, PA: U.S. Army Strategic Studies Institute, 1999.

Strausz-Hupé, Robert. *Geopolitics.* New York: Putnam's Sons, 1942.

Stripling, George. *The Ottoman Turks and the Arabs.* Philadelphia: Porcupine, 1977.

Subrahmanyam, Sanjay. "Connected Histories: Notes towards a Reconfiguration of Early Modern Eurasia." *Modern Asian Studies* 31, no. 3 (1997): 735–62.

Sumida, Jon. "Alfred Thayer Mahan, Geopolitician." *Journal of Strategic Studies* 22, nos. 2–3 (1999): 39–61.

———. *Inventing Grand Strategy and Teaching Command.* Baltimore: Johns Hopkins University Press, 1997.

Suzuki, Chusei. "China's Relations with Inner Asia." In *The Chinese World Order,* edited by John K. Fairbank, 180–97. Cambridge, MA: Harvard University Press, 1968.

Swaine, Michael D., and Ashley J. Tellis. *Interpreting China's Grand Strategy: Past, Present, and Future.* Santa Monica, CA: Rand, 2000.

Swoboda, Wincenty. *Warna 1444.* Cracow: Krajowa Agencja Wydawnicza, 1994.

Tamborra, Angelo. "Dopo Lepanto: Lo spostamento della lotta antiturca sul fronte terrestre." In *Il Mediterraneo nella seconda metà del '500 alla luce di Lepanto,* edited by Gino Benzoni, 371–91. Florence: Leo S. Olschki Editore, 1974.

Tétreault, Mary Ann. "Autonomy, Necessity, and the Small State: Ruling Kuwait in the Twentieth Century." *International Organization* 45, no. 4 (1991): 565–91.

———. *Revolution in the World Petroleum Market.* Westport, CT: Quorum Books, 1985.

Thayer, William Roscoe. *A Short History of Venice.* London: Macmillan, 1905.

Thompson, Laurence G. "The Junk Passage across the Taiwan Strait: Two Early Chinese Accounts." *Harvard Journal of Asiatic Studies* 28 (1968): 170–94.

Thompson, William. "The Military Superiority Thesis and the Ascendancy of Western Eurasia in the World System." *Journal of World History* 10, no. 1 (1999): 143–78.

Thompson, William, and Gary Zuk. "World Power and the Strategic Trap of Territorial Commitments." *International Studies Quarterly* 30, no. 3 (1986): 249–67.

Thuasne, Louis. *Gentile Bellini et Sultan Mohammed II.* Paris: Ernest Leroux, 1888.

Thucydides. *The Landmark Thucydides.* Edited by Robert B. Strassler. New York: Free Press, 1996.

Tilly, Charles. *Coercion, Capital, and European States, AD 990–1992.* Oxford: Blackwell, 1990.

———. *The Formation of Nation States in Western Europe.* Princeton, NJ: Princeton University Press, 1975.

Trevor-Roper, H. R. "Fernand Braudel, the *Annales,* and the Mediterranean." *Journal of Modern History* 44, no. 4 (1972): 468–79.

Turner, Ralph. "Technology and Geopolitics." *Military Affairs* 7, no. 1 (1943): 5–15.

U.S. Congress. Office of Technology Assessment. *An Assessment of Maritime Technology and Trade.* OTA-O-220. Washington, DC., October 1983.

U.S. Department of Defense. *Annual Report on the Military Power of the People's Republic of China.* FY04 Report to Congress on PRC Military Power. www.defenselink.mil/pubs/d20040528PRC.pdf.

———. *Quadrennial Defense Review Report.* Washington, DC, 30 September 2001. www.defenselink.mil/pubs/qdr2001.pdf.

U.S. Department of Transportation. Bureau of Transportation Statistics. *G-7 Countries: Transportation Highlights.* BTS99-01. Washington, DC, November 1999.

———. *Transportation Statistics Annual Report 2001.* BTS02-07. Washington, DC, 2002.

U.S. Department of Transportation. Bureau of Transportation Statistics. Maritime Administration. U.S. Coast Guard. *Maritime Trade and Transportation, 1999.* BTS99-02. Washington, DC, 1999.

U.S. National Intelligence Council. *Mapping the Future: A Report of the National Intelligence Council's 2020 Project.* NIC 2004-13. December 2004. www.cia.gov/nic/NIC_global trend2020.html.

Vagts, Alfred. "Geography in War and Geopolitics." *Military Affairs* 7, no. 2 (1943): 79–88.

Valencia, Mark. *The Proliferation Security Initiative: Making Waves in Asia.* Adelphi Papers, 376. London: International Institute for Strategic Studies, 2005.

Van Creveld, Martin. *Technology and War.* New York: Free Press, 1989.

van der Wee, Hermann. "European Long-Distance Trade, 1350–1750." In *The Rise of Merchant Empires,* edited by James D. Tracy, 14–33. Cambridge: Cambridge University Press, 1990.

Van Evera, Stephen. "Offense, Defense, and the Causes of War." *International Security* 22, no. 4 (1998): 5–43.

Veinstein, Giles. "Commercial Relations between India and the Ottoman Empire (Late Fifteenth to Late Eighteenth Centuries): A Few Notes and Hypotheses." In *Merchants, Companies, and Trade,* edited by Sushil Chaudhury and Michel Morineau, 95–115. Cambridge: Cambridge University Press, 1999.

Venables, Anthony. "The Assessment: Trade and Location." *Oxford Review of Economic Policy* 14, no. 2 (1998): 1–6.

Verlinden, Charles. "Venise entre Méditerranée et Atlantique." In *Venezia, centro di mediazione tra Oriente e Occidente,* edited by H.-G. Beck, M. Manoussacas, and A. Pertusi, 1:51–55. 2 vols. Florence: Leo S. Olschki Editore, 1977.

Vries, P. H. H. "Are Coal and Colonies Really Crucial? Kenneth Pomeranz and the Great Divergence." *Journal of World History* 12, no. 2 (2001): 407–46.

Walder, David. *The Short Victorious War.* London: Hutchison, 1973.

Waldron, Arthur. "Chinese Strategy, Fourteenth to Seventeenth Centuries." In *The Making of Strategy,* edited by Williamson Murray, Macgregor Knox, and Alvin Bernstein, 85–114. Cambridge: Cambridge University Press, 1994.

———. "The Problem of the Great Wall of China." *Harvard Journal of Asiatic Studies* 43, no. 2 (1983): 643–63.

Walt, Stephen. *The Origins of Alliances.* Ithaca, NY: Cornell University Press, 1987.

Waltz, Kenneth. *Theory of International Politics.* Reading, PA: Addison-Wesley, 1979.

White, John A. *The Diplomacy of the Russo-Japanese War.* Princeton, NJ: Princeton University Press, 1964.

Whittlesey, Derwent. *German Strategy of World Conquest.* New York: Farrar & Rinehart, 1942.

———. "Haushofer: The Geopoliticians." In *Makers of Modern Strategy,* edited by Edward Mead Earle, 388–411. Princeton, NJ: Princeton University Press, 1943.

Wills, John. "Maritime Asia, 1500–1800: The Interactive Emergence of European Domination." *American Historical Review* 98, no. 1 (1993): 83–105.

———. "Maritime China from Wang Chih to Shih Lang." In *From Ming to Ch'ing: Conquest,*

Region, and Continuity in Seventeenth-Century China, edited by Jonathan Spence and John Wills, 201–38. New Haven, CT: Yale University Press, 1979.

———. "Relations with Maritime Europeans, 1514–1662." In *The Cambridge History of China,* vol. 8, *The Ming Dynasty, 1368–1644,* edited by Frederick W. Mote and Denis Twitchett, pt. 2:333–75. New York: Cambridge University Press, 1998.

Wittek, John W. "Review Article: The Seventeenth-Century European Advance into Asia." *Journal of Asian Studies* 53, no. 3 (1994): 867–80.

Wittek, Paul. *The Rise of the Ottoman Empire.* London: Royal Asiatic Society, 1938.

Wohlstetter, Albert. "Illusions of Distance." *Foreign Affairs* 45, no. 2 (1968): 242–55.

Wójcik, Zbigniew. *Jan Sobieski (1629–1696).* Warsaw: Państwowy Instytut Wydawniczy, 1983.

Wolfers, Arnold. *Discord and Collaboration.* Baltimore: Johns Hopkins Press, 1962.

———. "The Pole of Power and the Pole of Indifference." *World Politics* 4, no. 1 (1951): 39–63.

Wolff-Poweska, Anna. *Doktryna geopolityki w Niemczech.* Poznan, Poland: Instytut Zachodni, 1979.

Woodward, C. Vann. "The Age of Reinterpretation." *American Historical Review* 66, no. 1 (1960): 1–19.

Wright, Gavin. "The Origins of American Industrial Success, 1879–1940." *American Economic Review* 80, no. 4 (1990): 651–68.

Wright, Quincy. *The Study of International Relations.* New York: Appleton-Century-Crofts, 1955.

Yergin, Daniel. *The Prize.* New York: Free Press, 1993.

Yost, David. "Political Philosophy and the Theory of International Relations." *International Affairs* 70 (April 1994): 263–90.

Zakaria, Fareed. *From Wealth to Power.* Princeton, NJ: Princeton University Press, 1998.

Zakythinos, Denis. "L'attitude de Venise face au déclin et à la chute de Constantinople." In *Venezia, centro di mediazione tra Oriente e Occidente,* edited by H-G. Beck, M. Manoussacas, and A. Pertusi, 1:61–76. 2 vols. Firenze: Leo S. Olschki Editore, 1977.

Lightning Source UK Ltd.
Milton Keynes UK
UKOW050334020212

186502UK00001B/39/P